W9-AQA-132

*f*P

Choke

WHAT THE SECRETS OF THE BRAIN REVEAL ABOUT GETTING IT RIGHT WHEN YOU HAVE TO

SIAN BEILOCK, PH.D.

FREE PRESS
New York London Toronto Sydney

Free Press
A Division of Simon & Schuster, Inc.
1230 Avenue of the Americas
New York, NY 10020

First Free Press hardcover edition September 2010

FREE PRESS and colophon are trademarks of Simon & Schuster, Inc.

For information about special discounts for bulk purchases, please contact Simon & Schuster Special Sales at 1-866-506-1949 or business@simonandschuster.com.

The Simon & Schuster Speakers Bureau can bring authors to your live event. For more information or to book an event, contact the Simon & Schuster Speakers Bureau at 1-866-248-3049 or visit our website at www.simonspeakers.com.

Designed by Mspace/Maura Fadden Rosenthal

Manufactured in the United States of America

10 9 8 7 6 5 4 3 2 1

Library of Congress Cataloging-in-Publication Data
Beilock, Sian.
Choke : what the secrets of the brain reveal about getting it right when you have to / Sian Beilock.
p. cm.
Includes index.
1. Failure (Psychology) 2. Performance—Psychological aspects. 3. Success—Psychological aspects. I. Title.
BF575.F14B45 2010
153.9—dc22 2010010595

ISBN 978-1-4165-9617-2
ISBN 978-1-4391-0962-5 (ebook)

NOTE TO READERS
Names and identifying details of some of the people the author knows and has worked with have been changed.

To my grandmothers,

Phyllis Beilock and Sylvia Elber,

each of whom modeled a spark and drive in her life's pursuits.

CONTENTS

Contents

INTRODUCTION

Ever since I was young I have been intrigued by amazing performances—at the Olympics, in the orchestra pit, and even my friend Abby's performance on the LSAT. How do people go about turning it on when it counts the most? Why do some thrive while others falter when the stakes are high and everyone is focused on their every move? As we know, sometimes that one instance of performance—one race, one test, one presentation—can change an entire life or a career trajectory forever.

My friend Abby and I have known each other since we were both thrown in the same dorm room freshman year at the University of California, San Diego. Although Abby and I shared a love for many things—the ocean, the Grateful Dead, and sappy movies—when it came to school, we couldn't have been more different. Throughout college I was constantly in the library studying for midterms and finals, writing papers, and rereading my notes from class. Abby was not. Now don't get me wrong, Abby did well in school, but you were more likely to find her at the beach than at the library and the likelihood that she would be daydreaming in class far outweighed the probability that she was actually listening to the professor lecturing in front of her. What amazed me most about Abby was her ability to perform well when the stakes were high. Abby wrote most of her English papers at four o'clock in the morning the night before they were due and reliably got A's on them, and those all-nighters in the library before finals always seemed to pay off for her.

After college, Abby decided to go to law school so she took the LSAT (the law school assessment test) and received a near perfect score. Abby took several steps to prepare for the big testing day. She bought a test-prep book and learned all the tricks of multiple-choice testing and she took practice after practice test to try to improve her score. By the time the real test day arrived, Abby was scoring in the upper quartile of all LSAT test takers, but her practice scores were nowhere near what she was able to pull off in the actual exam. Abby turned it on when it counted the most and her high performance made all the difference. In part because of that one day, that one four-hour test, Abby was admitted to the top law school in the country, had a leading firm recruiting her by the end of her first year, and landed a high-paying job when she graduated—a job that would never have been available to Abby if her performance on the LSAT had gone awry. One four-hour testing period, one-sixth of a day, changed Abby's life forever.

Psychologists are often accused of doing "me-search," that is, trying to understand themselves rather than "re-searching" others, and, admittedly, this holds for me as well. As a child, and even into my adulthood, I performed well on the sports field and in the classroom, but in certain situations I didn't always attain the high-level performance I was striving for. I had one of the worst soccer games of my life playing in front of college recruiters and I could never manage to score as well on the actual SAT as I did on the many practice tests I took before the exam. Abby faced similar high-stakes situations, yet the pressures didn't seem to faze her. Instead she thrived under them.

By the time I got to college I was hooked on figuring out why folks sometimes fail to perform at their best when the stakes are highest. I majored in cognitive science and soaked up as much as I could about how the workings of the brain drive learning and performance. But I always felt as if I was only getting half the picture.

As fascinated as I was by research on how we acquire skills such as language and math, I seldom came across people who were studying how the stresses of an important testing situation—for example, sitting for the SAT or ACT—might interfere with students' ability to show what they are made of. Perhaps because I split my time in college between the lacrosse field and the classroom, I also wondered how my academic ability was linked to my athletic prowess. Was my nervousness before a final exam related to the pressures I felt to make the big play in the lacrosse finals? If you are the kind of person who tends to bomb big tests, does this mean that you have a high likelihood of missing the game-winning shot as well?

These issues have tantalized me since I started school, stepped onto the playing field, held a musical instrument in my hand for the first time, and watched Abby ace every test. Yet I couldn't begin to find the answers until I went off to graduate school at Michigan State University, which provided me with the opportunity to work with professors doing seminal work in sport science, psychology, and neuroscience. Everyone thought I was crazy to move from the San Diego beaches to the snow, but my MSU education was unique in that I was able to learn about how the brain supports success in diverse performance arenas. Regardless of whether I was studying the complex decision-making processes involved in flying an airplane or how different parts of the brain work together to do math, my question about human performance was always the same: Why do we sometimes fail to perform at our best when it counts the most?

Early in my Ph.D. career, I convinced one of my advisers, Dr. Thomas Carr, to let me set up a putting green in his laboratory. We reasoned that if we could gain understanding into why golfers sometimes miss easy putts when there is a lot riding on their success, we might not only learn something about failure in sports, but we might also find out something interesting about why people make silly mistakes when taking a pressure-filled math test. After all, both golf and math are complex activities that take time and care to learn. And in fact we learned that, while poor performance under pressure was common in both tasks, their practitioners messed up their performances in different ways. To paraphrase Tolstoy, all unhappy performances resemble each other, but each is messed up in its own way.

Today, with the advent of new brain-imaging techniques, we can look inside the heads of players, students, and even people in the business world and make reasoned guesses about the types of programs that the brain is running. We are also able to get a handle on why these internal programs fail when people are under pressures that lead them to choke. In the past few decades, I've found answers to some of my nagging questions about human performance. These answers will change how you think about learning, assessments of intelligence, and talent identification—from the playing field to the classroom to the boardroom and beyond.

In *Choke* I present the latest information on what we psychologists know about how people learn and perform complex skills. I address questions that include: What are the brain systems that oversee how we pick up sports skills? Does the way we develop sports skills really differ from the ways we learn in the classroom or perform in the orchestra pit? How do we flub performances in these different settings? Why do some people fail while others succeed when everything is riding on their next move and the pressure to excel is at a maximum?

When I arrive in my office Monday morning, it's not unusual for me to find several phone messages from parents who want to know why their child plays well in practice during the week, but fails during competition over the weekend, or from high school students who are interested in ensuring that the high test scores they're receiving on their practice SATs will translate to the actual exam. I am intrigued by each and every case because it's only by understanding how less-than-stellar performances come about that we can create the right strategies to ensure that we can succeed when it counts the most.

I give a good number of lectures for corporations every year in which I shed light on what brain science says about how to perform at our best in the heat of negotiation or when a crisis strikes. My hunch is that one of the reasons companies are eager to have me speak is that it's hard to put a finger on exactly why unexpected failures occur when the stakes are high. *Choke* will change this.

As a society, we are obsessed with success, and because of this, people are constantly trying to uncover the ingredients that produce amazing performances. The flip side of the coin of success, of course, is failure. And uncovering the mechanism by which you flub an important sales pitch or bomb a negotiation provides clues for how you can perform your best in any situation.

You've no doubt heard the phrase "choking under pressure" before. People talk about the "bricks" in basketball when the game-winning free throw devolves into an airball, the "yips" in golf when an easy putt to win the tournament stops short, and "cracking" in important test-taking situations when a course grade or college admission is on the line. Others talk about "panicking" when someone isn't able to think clearly enough to follow well-practiced procedures to exit a building in a fire. But what do these terms really mean?

Malcolm Gladwell, in his 2000 *New Yorker* essay "The Art of Failure," talks about choking and panicking. The former, Gladwell suggests, occurs when people lose their instinct and think too much about what they are doing. Panicking occurs when people rely on instincts they should avoid. I am here to tell you that from a scientific point of view these are both instances of choking.

Choking can occur when people think too much about activities that are usually automatic. This is called "paralysis by analysis." By contrast, people also choke when they are not devoting enough attention to what they are doing and rely on simple or incorrect routines. In *Choke* you will learn about what influences poor performance under pressure in myriad situations so that you can prevent failure in your own endeavors.

But first, what is choking exactly? Choking under pressure is poor performance that occurs in response to the perceived stress of a situation. Choking is not simply poor performance, however. Choking is suboptimal performance. It's when you—or an individual athlete, actor, musician, or student—perform worse than expected given what you are capable of doing, and worse than what you have done in the past. This less-than-optimal performance doesn't merely reflect a random fluctuation in skill level—we all have performance ups and downs. This choke occurs in response to a highly stressful situation.

A business executive recently shared a story with me about an incident at his company that occurred shortly after the big anthrax scares of 2001. Letters containing anthrax spores had been mailed to news media and political figures, resulting in several injuries and even five deaths. First tested as a biological warfare agent in the 1930s, anthrax spores are easily spread, so companies developed procedures after the mailings for employees to follow in order to contain contamination if it were to occur. This executive's company had developed a step-by-step guide for employees who suspected they had been exposed and had even conducted several company-wide practice drills to go over procedures in the event of an anthrax contamination. Nonetheless, when one day white powder came spilling out of a package that a woman in his division had opened, instead of calmly following procedure, which first and foremost meant staying put, she immediately panicked and ran out of the office, making contact with several employees along the way.

Fortunately, the incident turned out be a hoax. What the executive wanted to know was whether this woman's panic was similar to an athlete faltering at the Olympics or his son freezing at the chalkboard in school. If so, then perhaps he could learn something from these activities about how to prevent panic in his own employees. These are all instances of choking under pressure. Knowing how particular chokes are similar (and different) is the key to treating them.

In *Choke* we will explore how performance in the classroom is tied to performance on the basketball court or orchestra pit and whether success in one arena carries implications for skill execution in another. We will ask why the mere mention of differences between the sexes in math ability disrupts the quantitative exam performance of a female test taker and we will delve into other activities where similar phenomena occur. Why are those high-powered students—with the most knowledge and skill—*most* likely to choke under the pressure of a big exam? Do these same folks also choke in sports? Can calling a "time-out" immediately before the game-winning field goal in football reduce a kicker's success or "ice" him? Why does *icing* work and can a politician be iced before giving an important speech? *Choke* tells the stories of the science behind these human performances and others as it explains what the secrets of the brain can teach us about our own success and failure at work and at play.

CHAPTER ONE

THE CURSE OF EXPERTISE

Russian tennis player Dinara Safina has had quite a bit of success on the world tennis scene. The twenty-three-year-old has competed in two Grand Slam quarterfinal matches, two semifinal matches, and three finals. She even spent time as the World Tennis Association's number-one-ranked player in 2009. But despite seven advanced round appearances, Safina has yet to win one of professional tennis's Grand Slam titles. Any way you slice it, Dinara Safina has faltered while playing on the biggest tennis stages.

It's crucial to a tennis player's reputation to win championship matches. Even though you may be the top player in the world according to the computers, if you can't prove your worth when it counts the most, you become known as a "choker." And when you show that you are not a clutch player, people stop rooting for you. Even you stop rooting for yourself.

"I put too much pressure on myself," Safina remarked after a surprising early third-round exit at the 2009 U.S. Open. Now, every time Dinara Safina steps on the court, whether or not she can win a big title is what's on her (and everyone else's) mind.

My job is to help people avoid these types of failures.

Today, I've been booked to speak at a Fortune 500 company retreat at Robert Redford's posh Sundance Resort in the mountains of Utah. The company president has gathered all her VPs for two days of brainstorming about ways to identify the best and brightest hires and how to help these hires perform at their best under the most pressure-filled circumstances. Basically, I have come to Utah to help the VPs prevent their employees (and themselves) from going down Dinara Safina's path.

I had spoken to the president over the phone a few days before my trip and, from our chat, had gotten the impression that she intended the weekend to be about 90 percent work and 10 percent rest and relaxation. Now, as I look at her VPs sitting slightly slouched in their chairs in front of me, and outfitted in shorts, sandals, and T-shirts, I get the feeling that they may have reversed the ratio of work to play. Nonetheless, they are all gathered in the resort's main conference room to hear what I have to say.

I have a hunch that my audience will be a bit skeptical of whether I, a psychology professor at the University of Chicago, have anything useful to report. After all, what does an academic know about the business world? These VPs are not particularly interested in psychology, my field of study, but they are interested in what it takes to be successful. Fortunately, I happen to know something about this because I study—to put it simply—human performance. My research tries to explain how people succeed at what they do, whether it's on the playing field, the putting green, in the orchestra pit, or in the boardroom.

My work isn't limited to identifying the keys to success at work and at play. I also study why people fail to perform their best when the stakes are high and everything is riding on their next move. In my research I explore how a straight-A high school senior can score three hundred points lower on his actual SATs than he did on the practice tests he took repeatedly before the big testing day. I am also interested in how a professional golfer like Greg Norman could enter the last day of the 1996 Masters with a six-stroke lead and end up losing by five strokes. Or how Dinara Safina let another Grand Slam title slip through her fingers by double-faulting on match point at the 2009 French Open finals. Finally, I try to understand why some of the very team members who work under the VPs I am speaking to today might bomb their next big presentation in front of an important client. I want to know how we can identify talent, those most likely to fail and those most likely to succeed when the pressure is on.

Today I am speaking without a computer—with only a piece of paper with a few notes jotted down to guide me. Even though I give several talks each year, it can be a bit daunting to be without the security blanket of my computer and the accompanying graphs of research data to guide my speech, especially when I'm not talking to college students who at least have to feign some interest in what I say to earn a grade. Because I research why people fail when they are under stress, if I find myself at a loss for words, I can always joke about doing "me-search," researching the types of pressure-induced failures that I, myself, have been known to fall prey to.

As the room quiets down, I explain that my goal for the next few hours is to provide the science behind the intangible: creativity, intelligence, choking under pressure. Anyone can become a better leader, worker, and performer with the realization that seemingly trivial aspects of the environment and of their mind-set can greatly affect success. They don't yet look convinced, but everyone perks up when I start to unpack some pretty counterintuitive research findings that just might change how they operate. For instance, the more experience you have as a leader of your company, the worse your ability to manage your team members can become. More experience, worse performance. This seems crazy, but I can back up my claim with actual research.

EXPERTS COULD USE A CRYSTAL BALL

Before becoming a professor in the Department of Management Science and Engineering at Stanford University, Pamela Hinds spent several years working at Pacific Bell and Hewlett-Packard trying to figure out how people's work and daily lives change with the introduction of new technology like computers and cell phones. Ten years ago it was hard to imagine that almost everyone would be reachable by cell phone. Part of Hinds's job was to anticipate these digital advances and determine how they would affect people's work, life, and play. Today, in her academic position, Hinds pursues many of the same research questions she did when she was in the business world. One topic she's been exploring is how people who develop and market new digital technology estimate the time and difficulty it will take others to learn to work these fancy tools.

Most of us have had the frustrating experience of fumbling with a new cell phone or handheld device, wondering if the developers devoted even a minute of their time to making the use of our new toy anything less than agonizing. My friend Jackie counts herself as a digitally frustrated consumer. Several years ago, Jackie's law firm bought her and all the other lawyers new handheld digital organizers. "It's supposed to make my life easier?!" she skeptically exclaimed to me one day over coffee. Jackie had graduated at the top of her class from the University of California at Berkeley's Bolt Hall School of Law and was already earning accolades at her first job at a high-powered San Francisco law firm. But she was not what one would call "computer savvy" and, as she was forced to adopt more and more technology, her detestation for all things electronic was growing. The day after Jackie was given the handheld device, she sat down at her kitchen table to learn how to operate her new tool. A couple of hours later, she let out a big sigh and returned it to the box, where it has sat on her kitchen table for two years, never touched again.

Hinds's goal is to ensure that cases like Jackie's don't occur too often. One way to do this is to understand how digital technology developers anticipate the trouble that folks like Jackie will have with their new toys. If the highly knowledgeable makers of these products can foresee the problems that new users will have, then—hopefully—such problems can be eliminated.

Experts are called upon to predict the performance of those less skilled all the time. Managers like the VPs to whom I am speaking today, for example, must estimate how much time it will take their employees to complete a project. Teachers must predict whether their students will be able to complete homework assignments and tests in the allotted time. Baseball coaches must understand the types of problems that a pitcher may encounter when learning to throw a new curveball. If not, how will the coaches devise the right training techniques to help their pitcher out of trouble? Yet stepping outside your own point of view and relating to people who have less knowledge and skill is not such an easy task. Managers, teachers, and coaches are not as good as one might think at these estimates. Hinds's work has shown us that these experts make particular kinds of mistakes when anticipating the performance of beginners.

Because Hinds spent several years working at Pacific Bell, it isn't so surprising that she decided to use cell phones to explore how experienced individuals go about understanding the performance of novices. Hinds asked sales clerks, cell phone customers (with some experience using the phones), and other people (with no experience at all with the new technology) to estimate how long it would take a new user to master the phones.[1] Yes, cell phone novices do exist—or at least they did in the mid-1990s, when this study was conducted.

After everyone had given his or her learning estimates, Hinds asked the folks who had never used a cell phone before to stick around and actually learn how to operate the new technology. The new users only had the written instructions that come tucked in the phone's package as a guide. Hinds kept track of how long it took these people to store a greeting on voice mail, for instance, or to listen to the messages in their voice mailbox. She then compared the novices' learning time with the estimates that everyone had given for how long it would take a new user to master the phone.

You might expect that salespeople would be stellar predictors. After all, they are experts in the technology they are trying to sell and they also deal with ignorant customers on a daily basis. Surely they must have a good understanding of the problems that new users face and of the time it will take them to succeed. But, unfortunately for the salespeople, something gets in the way of experts' ability to predict novices' performance. We psychologists call this something the *curse of expertise.* There is no crystal ball.

Salespeople focused so closely on their own performance and how effortlessly they operated the phone that they had a hard time predicting novices' mistakes. Because of this, salespeople were the least accurate predictors of the new users' learning time. It took the new cell phone users around thirty minutes to complete all the tasks—such as recording an outgoing voice-mail message and saving and deleting mailbox greetings. The salespeople estimated that it would take novices less than thirteen minutes to master all of these phone skills. This is just about the time estimate given by the people who had never used the phones before. So, our experts performed like novices. They fell victims to the curse of expertise. Interestingly, the customers who had some (but not a ton) of cell phone experience were the most accurate predictors.

The "curse of expertise" trips up experts when they try to predict others' performance.

These findings puzzled Hinds, so she decided to give the salespeople some help. In another part of the experiment, before asking the sales men and women to predict new users' learning time, Hinds had the salespeople recall some of their own learning experiences—thinking back to the points of confusion and problems that they themselves had encountered when attempting to use the new phone for the first time. She even told the salespeople to keep these experiences in mind when making their predictions. Unfortunately, her advice didn't help. The salespeople again underestimated the time it would take novices to complete the phone tasks—not changing much from their original thirteen-minute estimates.

How could it be the case that experts perform like novices? Well, something interesting happens as people get better and better at performing a skill—and this occurs whether we are talking about programming a cell phone, riding a bike, or parallel parking in city traffic. They forget stuff. Think about riding a bike. How exactly do you do this? Well, yes, first you have to get on the bike and pedal. But there is a lot more to it than that. You have to balance, hold on to the handlebars, look at what is in front of you. If you miss any of these steps, falling is a real possibility. This usually doesn't happen when proficient bike riders are actually riding, but if you were to ask a bike rider to explain the "how-tos" of this complex skill, he would forget details. This is because the proficient bike rider is trying to remember information about bike riding that is kept as a *procedural memory*, as we psychologists term it.

Procedural memory is implicit or unconscious. Although procedural memory is mostly used to guide the performance of athletic skills, it creeps up in all sorts of tasks—like programming a cell phone. Similar to running a play in football or acing your opponent in tennis, operating a cell phone (for example, navigating through several screens to the point where you can enter a password to retrieve your voice messages) involves complex motor movements linked together to reach a goal.

You can think of procedural memory as your cognitive toolbox that contains a recipe that, if followed, will produce a successful bike ride, golf putt, baseball swing, or fully operating cell phone. Interestingly, these recipes operate largely outside of your conscious awareness. Your own facility is because, when you are good at performing a skill, you do it too quickly to monitor it consciously. This makes it hard for you to articulate what is in your procedural memory. If you don't think about the specific steps you go through while performing a task, reporting these steps to someone else (or using these steps to estimate the time it might take others to perform the same task) can be difficult.

Procedural memory is often distinguished from another form of memory: our explicit memory that supports our ability to reason on the spot or to recall the exact details of a conversation we had with our spouse the week before.[2] It might seem odd that we have some memories in our heads that we can't actually get at by looking around for them and others that we can consciously access, but once you know something about the makeup of the brain, this memory distinction is not so surprising. Simply put, explicit and procedural memories are largely housed in different parts of the brain and some activities rely more on the former type of memory and some on the latter.

Perhaps the strongest evidence for this memory division comes from the case of Henry Gustav Molaison, or H.M. On September 1, 1953, H.M. had surgery that was designed to remove a portion of the temporal lobes in his brain (specifically, the hippocampus) in an attempt to put an end to epilepsy-related seizures that couldn't be controlled by medication. Although H.M.'s seizure disorder was eradicated by the surgery, he lost a majority of his hippocampus in the process, the brain structure that is involved in transferring new information we encounter into explicit, long-lasting memories. As a result, H.M. lost his ability to create new memories that lasted for more than a few seconds. If you would see him again after meeting him only a week before, he wouldn't remember you. Interestingly, H.M. could still learn skills (such as the sequencing of finger movements needed to play the piano or how to trace a figure by looking at its reflection in the mirror) that rely heavily on procedural memory because this memory resides in brain areas, such as the motor cortex, basal ganglia, and the parietal lobe, that were not removed in the surgery.[3] Of course, when asked, H.M. couldn't tell you in detail how he performed activities based on these procedural memories, just as people with intact brains can't tell you. Just think about the type of description you might get from Michael Jordan if he were asked how he dunks a basketball. He might invoke the Nike motto and say that he "just does it," not because he doesn't want to give away his flying secrets, but because he may not know what he does. As we get better and better at performing skills such as operating a cell phone or riding a bike, our conscious memory for how we do it gets worse and worse. We become more expert and our procedural memory grows, but we may not be able to communicate our understanding or help others learn that skill.

> As we get better at performing a skill, our conscious memory for how we do it gets worse and worse.

At this point in my talk, the man sitting in front of me interrupts, introduces himself as John, and offers to share a story. I encourage him to go on and John recounts something that had happened to him a few months ago when his IT team was in the process of proposing a major change in the online flight reservation program used by one of the big airline companies—a change that would attract more customers to the airline's Web-based reservation system and make the customers' experience more enjoyable once they got there.

John had thrown out a deadline to his employees for developing the new software and for coming up with a presentation plan for the clients who would be in the office the following week. John did this on a Monday and had expected to see his team in his office by the end of the day Friday with the beginning of tangible products in their hands. All day went by on Friday and not one of his team members came by. Finally, toward the end of the day, one of his middle managers came to him and boldly announced that they didn't have enough time to complete their tasks. At first John was quite annoyed. But after the employee explained how many hours they had been working and described all of the technical issues they had come up against, it dawned on John that he had underestimated the complexities of the work his team members had to complete. John himself had dealt with these complexities in the past when he was in his employees' position, but he had totally forgotten about them when estimating how quickly his team could pull together the new product.

So what can business managers do to become better estimators of their team members' skill and ability? Consulting with someone less experienced might do the trick. Recall that in Hinds's research, the customers with *some* cell phone experience were the most accurate predictors of the time it would take a new user to master the phones. John had begun employing a similar strategy with his team that seemed to be working. Before handing out a big project, he now surveys several of his employees to get an idea of the problems they think they might encounter and the time and support they anticipate needing in order to resolve them. John thinks that asking these questions beforehand has helped put him and his employees on the same page. This leads to more accurate work estimates for his clients, on-time performance, and happier customers overall.

Experienced people benefit from hearing the thoughts of those who are less skilled in areas other than the business world. Pairing two university students of lower and higher math ability together, for example, leads to better performance on tricky math problems than either student could produce alone. It's not surprising that students who are lower in math ability benefit from the guidance of stronger math students, but it is interesting that stronger math students also benefit from these pairings. This is because, when you have to teach someone who knows less than you, you end up learning the material better yourself. Poorer students can also help the stronger students to think about a problem differently or "outside the box," which facilitates the type of creativity that is often needed to solve atypical problems in new, intuitive ways. Sometimes it's a good idea for those with experience to enlist the help of the less knowledgeable.

Experienced people benefit from hearing the thoughts of those who are less skilled.

Even when no one knows the answer to a difficult problem, several heads are often better than one. When biology students at the University of Colorado answered questions in class using clickers (the handheld device that allows instructors to survey students during a lecture), discussed the question with their neighbors, and then revoted on the same question, the percentage of correct answers increased.[4] This was true even when none of the students in the discussion group originally knew the answer. Discussing a problem with others brings alternative problem perspectives together and, more often than not, this type of communication leads to the optimal solution. As one student commented, "Discussion is productive when people do not know the answers because you explore all the options and eliminate the ones you know can't be correct." Justifying an explanation to another person also provides people with valuable opportunities for developing communication and reasoning skills. This is the case whether you are new to an area or have tons and tons of experience.

CHOKING UNDER PRESSURE: WHY, WHEN, HOW

Now that I have talked about some of the limitations of expertise, I should tell my audience more about why highly skilled individuals may not always perform at top capacity and what can be done about it. During my phone call with the company president the week before, she had mentioned that her VPs spend a lot of time giving (and preparing their team members to give) presentations to their clients in what she referred to as "do-or-die" situations. If the presentation goes off without a hitch, and especially if the client's on-the-spot questions are answered satisfactorily, they get the account. If not, money, clients, and future opportunities walk out the door. Second chances do not occur too often in their business.

As I was listening to the president, I couldn't help but think that this type of pressure sounded similar to that which a high school senior feels as he goes in to take the all-important SAT, and a golfer experiences as she attempts to sink the putt needed to make the LPGA tour cutoff, and even a violinist auditioning for first chair in a major symphony orchestra undergoes as he prepares for his solo. If these people perform at their best, doors open. If they perform badly, well, there may not be a second chance.

Of course, sometimes even having a second chance doesn't help. Think about the lovable American ice skater Michelle Kwan. Though she was commonly considered to be the world's best figure skater in the late 1990s and early 2000s, she was never able to clinch the all-around Olympic gold. In the 1998 Winter Olympics, Kwan was upset by American teenager, Tara Lipinski. In her second chance at Olympic gold in 2002, Kwan led after the short program, followed by another American teen, Sarah Hughes. Kwan seemed a shoe-in, but in the final free skate, Kwan looked visibly stiff, two-footed a combination, and fell on her triple flip. Hughes, on the other hand, pulled off a perfect performance and won the gold. Some people, even when given two chances, can't perform up to their potential.

Millions of fans were dying to see Kwan win. When stakes are so high, even highly experienced performers feel pressure to succeed.

Although Sarah Hughes seemed to be energized by the importance of what she was doing and managed to perform at her best, this same pressure caused Kwan to perform at her worst. Improved performance when it counts the most is understandable—people may simply put a little more effort into what they are doing because they are motivated to succeed. But failure to perform under pressure is a phenomenon in need of an explanation. This is where I come in.

In my Human Performance Laboratory, at the University of Chicago, we study people in high-pressure athletic, academic, and business situations and we do so with one goal in mind: to understand why, when, and how failure under pressure occurs. Our belief is that if we can get a handle on why people fail when the pressure is on—in other words, why people *choke under pressure*—then we can develop strategies to alleviate it.

So, why do some people shine while others fail when it counts the most—whether it be on a high-stakes college admissions test or on the Olympic stage? Why, at the 2002 Olympics, did Michelle Kwan fall while Sarah Hughes landed perfectly on every jump? Are all types of pressures the same? What can you do in your own job or field of performance if you happen to find yourself falling rather than sticking the landing under pressure?

In one sense, the pressures that come with sitting for a high-stakes test, pitching to a client, or stepping on the Olympic stage are very similar. People have a desire to succeed and this can, ironically, cause them to do their worst. But how exactly pressure causes a performance hiccup depends on *what* we are doing and the type of memory that is driving how we execute that particular skill.

As I have written above, our memory (and the tasks we engage in) can roughly be divided into explicit and procedural forms. In the former case you have activities such as adding numbers in your head, reasoning through a difficult issue with your client, or recalling what was said in a heated argument you had last week with a coworker. In the latter case, it's taking a golf swing, landing a double axel on skates, or operating a cell phone. Because different skills rely on different types of memories, the answer to questions about why people fail to perform at their best and what can be done to prevent it is not "one size fits all."

I decide to start by providing my audience with information about choking in academic settings. Any insight we can make into why people choke will of course advance our understanding of human performance. Yet, I warn my audience that understanding the choking phenomenon and what specifically you can do to prevent it in your own career or field requires a look at research findings across many different performance arenas—from the classroom to the boardroom and beyond.

HIGH-POWERED FAILURE

Johann Carl Friedrich Gauss was a German scientist (1777–1855) and mathematician known for his role in number theory and statistics. A precocious teenager, he made many of his mathematical discoveries at an early age. In fact, at the age of twenty-four, Gauss published a book, *Disquisitiones Arithmeticae*, that presented some of the most brilliant mathematical theory of the time. Gauss was certainly an exceptional mathematician, but the reason I am most interested in him is that, in my lab, we teach students some of his math to get a handle on who will forget what they learned when they find themselves in a stressful exam.

One of the mathematical theories that Gauss developed was a system of arithmetic called modular arithmetic, or, as we call it in our lab, mod math. We researchers like this math task because most students have not seen it before. People *do* have experience with the types of calculations needed to solve mod math problems (just as they have experience with the types of problems seen on the SAT or GRE), but they *haven't* encountered the exact same problems in the past. This means that, if some choke while others sail, we know that these performance differences can't simply be due to who has seen mod math before. Everyone comes into our lab with a blank slate, in a sense, and we teach each student how to use basic math procedures to successfully solve mod math problems. Our goal admittedly is a bit perverse. We want to teach people to solve mod math so that we can see if they get worse when we put them under stress. But, in our defense, this perverse goal is needed if we truly want to get a handle on why they choke.

We usually teach people a two-step method to solve mod math problems such as: $32 \equiv 14 \pmod{6}$. First, you subtract the middle number from the first number: 32 minus 14. Then, you divide this answer by the mod number (here, 6). If the answer to $32 - 14$ is a multiple of 6, meaning that 6 goes into the answer of $32 - 14$ (that is, 18) evenly without a remainder, the equation is said to be "true." If not, it's "false." Another way to figure out the validity of a mod math problem is just to divide the first two numbers by the mod number. If, when divided by the mod number, both numbers have the same remainder (here, both 32 and 14 divided by 6 have a remainder of 2), then the equation is true.

Just as when people take the quantitative portion of standardized tests such as the GRE (the major test used in graduate school admissions), we present mod math problems to folks one at a time on the computer screen and ask them to solve the problems as quickly and accurately as they can. However, we are not so focused on how people perform at mod math when they have all the time in the world and there are negligible consequences associated with making errors. Rather, we want to know how people's math performance changes when we stress them out.

In one experiment that my graduate student Marci and I conducted, we brought around one hundred college students, one by one, into our lab to perform several dozen mod math problems.[5] For this particular experiment, Marci had put up fliers around campus offering money in exchange for volunteering for a psychology experiment about problem solving. We were careful not to mention anything specifically about math on our fliers because we didn't want to attract only people who like to do math. Instead, we wanted to get a variety of people into our lab so that we could see how different people might react to a stressful exam situation.

When students arrived for the experiment, Marci thanked them for coming, led them into one of our testing rooms, and seated them in front of a computer. Marci told the students that they would be performing the mod math task and explained how to solve it. Some people rolled their eyes or even groaned when they heard they were going to be doing math, but most were eager to get going. After students got some practice, the real experiment began. Everyone was, at first, simply told to do his or her best—to try to be as quick and accurate as he or she could when solving the problems. But then we made things a bit more interesting. Marci asked all the students to complete a second set of problems. This time, however, immediately before starting the math test, she mentioned a few things intended to up the ante for our students:

> So, we thought we would let you know that we have tested the problems you are about to do on other students last semester, and we got an average of how many they got right and how quickly they solved the problems. In the next set of problems, we will calculate a score for you based on these same two components—how quickly and how accurately you solve the problems. If you can do better than the average student from last year, then at the end of the experiment, we will give you twenty dollars.
>
> But there's a catch. What we are interested in in this experiment is teamwork and how people work together. So as part of this experiment you have been paired with another person. In order for you to get the twenty dollars, not only do you have to improve by twenty percent but the person you have been paired with has to improve as well. So it's a team effort. Now I can tell you that you are actually the second person in the pair—the person you have been paired with came in this morning and did improve by twenty percent. So if you can improve now, you will get the twenty dollars and so will they. But if you don't improve, they don't get the twenty dollars and neither do you. Do you have any questions?
>
> Also, your performance is going to be videotaped on these problems. Some professors and students here at the university, and math teachers in the area will be watching the tapes to see how people are performing. Okay, I am going to set up the video camera now and then we can begin.

As most of my audiences do, the VPs to whom I recounted our pressure manipulation cringed as they identified with the students we were testing. But, I told the VPs that, immediately after the students perform the math problems, we tell them that our pressure scenario is bogus and we give everyone the money—regardless of how they actually perform. You might imagine that students in our study would be upset about being deceived, but we explain to them that to study the impact of high-stakes testing situations, we must create a stressful environment in our laboratory. Only then can we develop test-taking and practice techniques to alleviate the ill effects of pressure. Most of the students who take part in our studies have performed under the gun at one time or another, so they are usually genuinely interested in the research and its findings. Of course, no student complains about walking out of our laboratory twenty dollars richer, either.

The types of pressure we use in our laboratory are common in real-world pressure cookers. The money we offer our students to succeed at mod math stands in for the scholarships they can earn if they do well in an actual testing situation or on the playing field. The threat of evaluation from reviewing the videotape also represents real-world assessment situations—just as SAT scores are judged by parents, teachers, and peers, and athletes' public performances are judged for medals at the Olympic Games.

Of course, the level of pressure we invoke in our laboratory doesn't even come close to the magnitude of stress that people feel in real do-or-die situations. Yet they still show some striking effects. In the study I have been recounting to the VPs, the math performance of the students we tested got worse when we put on the pressure. Most interesting, however, was who showed the biggest performance drops under stress.

In my audience of VPs, John—the manager who had come up with his own trick for accurately estimating his employees' performance needs—spoke up. "My daughter was the only student in her eighth-grade algebra class last year to receive perfect scores on every homework assignment she completed. Yet she could never pull off stellar performance on the exams. It was as if, when the pressure was on, she was unable to use what she knew to succeed. I bet it was the smartest kids in your study, the kids who may actually have known a lot, whose performance dropped the most under stress." Others around him shook their head, some in agreement and some skeptically. In fact, our findings revealed that John was, in a way, correct.

What I had not yet told my audience was that, in another study session, Marci and I had collected measures of all of our subjects' *working-memory* capacities. We will get into this concept in more detail in later chapters, but for now, just think about working-memory as your cognitive horsepower.

Rooted in the prefrontal cortex, working-memory reflects our ability to hold information in memory in the short term, but it's more than just storage on a computer's hard drive. Working-memory involves the ability to hold information in mind (and protect that information from disappearing) while doing something else at the same time. For instance, working-memory is in play when you are trying to remember the address of the restaurant you are heading to while at the same time reading a text from the friend you are meeting up with for dinner.

Why did we care about students' working-memory? Several studies have shown that working-memory differences across people account for between 50 percent to 70 percent of individual differences in abstract reasoning ability or fluid intelligence.[6] In short, working-memory is one of the major building blocks of IQ.

ASSESSING WORKING-MEMORY

The test Marci and I used to assess students' working-memory (aka cognitive horsepower) really gets at the ability to keep information safe in memory while a person may be distracted with some other task or goal. Similar to trying to remember a restaurant address and read a text, in our test people were asked to remember a list of letters while simultaneously reading a series of sentences aloud. When accurately assessing folks' general cognitive horsepower, it doesn't much matter

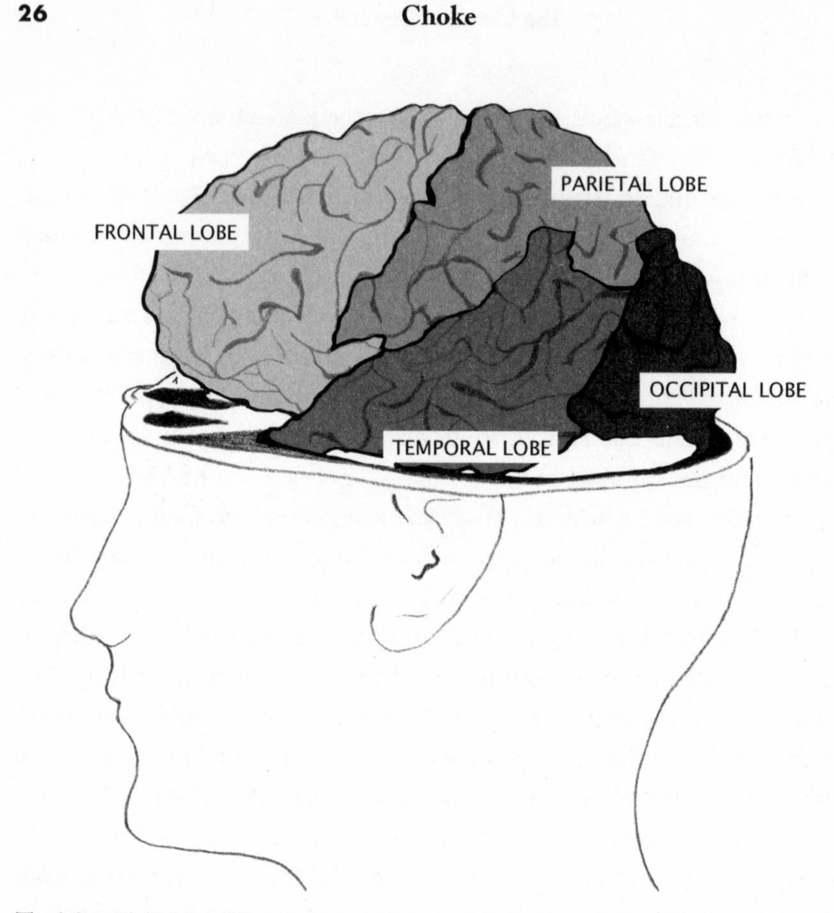

The lobes of the brain. The prefrontal cortex is the very front part of the brain housed in the frontal lobes.[7]

what people are holding in mind, just that we can measure their ability to shield this information from disappearing while they are doing something else at the same time.

Let me give you a bit more detailed example. In one task, called the Reading Span task,[8] a person might be asked to read—out loud—the following sentences one at a time on a computer:

On warm sunny afternoons, I like to walk in the woods. ? F

The farmer drove the grape to the sleeping bear. ? E

The ranger saw the eagle in the sky. ? D

The man thought the light was a nice after dinner train. ? R

After work, the woman always goes home for lunch. ? B

After reading each sentence, people are told to report whether the sentence makes sense and then to read the letter out loud at the end of the sentence. Then the sentence and letter disappear from the screen, and the next sentence-letter pair appears. Deciding whether the sentences are sensible is simple: the first sentence makes sense and the second one doesn't. But this part of the test is a red herring: We are not so interested in how test takers perform on the sensibility part of this task. What we want to capture is people's ability to remember the letters at the end. After a number of sentence-letter pairs (usually somewhere between three and five, we ask folks to recall all the letters that came at the end of each of the sentences they read and to do so in the same order they initially saw them (in our example, F, E, D, R, B). People know they are going to have to recall the letters from the outset, but they don't know *when* the recall task is coming. So people must maintain the letters at the end of the sentences in memory while making sensibility judgments about what they are reading. Holding information in memory while doing something else at the same time gets at the heart of working-memory.

Working-memory plays an important role in most of what we do on a daily basis. Remembering a phone number while taking a hot casserole out of the oven, planning the turn you have to make two streets down while weaving through traffic, or comparing in your mind's eye how your new sofa might look at different angles and configurations in the living room before it is actually moved, all involve working-memory. And the amount of working-memory that you have often predicts how well you will perform in activities imperative for academic success, such as reading comprehension or problem solving. So, you might be surprised to learn that it was our higher-powered students—the ones with the most working-memory—who performed the worst under pressure.

Not surprisingly, students with higher working-memory did out-

Higher-powered students—the ones with the most working-memory—fail under pressure.

perform everyone else by about 10 percent when the mod math problems were just for practice. But when the pressure was on, the performance of those highest in cognitive horsepower fell to the level of those who were lowest. The performance of individuals with lower working-memory didn't decline under pressure. Why?

To answer this question, Marci and I went back and took a closer look at our math problems. As you now know, in performing mod math, the task is to judge whether equations such as 32 ≡ 14 (mod 6) are true or false. Although you can solve these equations by going through several subtraction and division steps—first subtract 14 from 32 and then divide this answer (here, 18) by 6—you can also take shortcuts in finding an answer. For example, if a student decided that problems with all even numbers are probably true (because when dividing two even numbers there is usually no remainder), this shortcut will produce the correct answer on some trials: 32 ≡ 14 (mod 6), but not always: 52 ≡ 16 (mod 8). When people use a shortcut such as "if all the numbers in the problem are even, respond yes, if not, respond no," they get rid of the need to hold problem steps in memory. Instead they can arrive at an answer pretty effortlessly. But shortcuts like this are not always correct.

Marci and I found that students with higher working-memories were more likely to go through the subtraction and division steps to get to the right answer precisely because they have the cognitive horsepower to compute answers in this way—"if you've got it, flaunt it." Instead, students with lower working-memory were more likely to rely on simpler shortcuts.

Under low-stress conditions, using more brainpower gets you further. That is why higher-working-memory individuals perform the best in practice situations. Under pressure, however, a majority of our high-powered students panicked and actually switched to the shortcuts that the lower-powered students normally used. Low-powered students also panicked, but because their usual shortcuts don't require a lot of effort (remember, they are essentially no more than good guesses), they stuck with them and their performance didn't drop under stress.

I pause to throw out a few questions to my VP audience: How many of you have found yourselves (or seen your employees) go for an easy answer or quick fix when the pressure is on, just as the high-powered students in our math test switched to the easy way out? In setting up this seminar, their company president actually commented that their group could improve how they handle unexpected situations in times of crisis. Whether dealing with an unanticipated on-the-spot question from a client who needs a well-thought-out and reasoned answer or a last-minute tweak to a presentation being given later that afternoon, managers and employees suffer when under pressure. The company president thought that, if they could learn to take the time to stop, pause, and regroup when under fire, they might find a way out of sticky situations.

There is a good deal of evidence that this stalling strategy helps—at least when people are engaged in activities that depend on explicit memory and require a lot of working-memory or cognitive horsepower. Pausing in the middle of a challenge can prevent you from going down the wrong solution track. Of course, this isn't always the case. When what we are doing relies on procedural memory that operates largely outside of conscious awareness, too much time and concentration can be a bad thing because we are tempted to tinker with skills that are best run off without interruption (we will get into this issue in more detail in the sports and musical performance chapters to come). But when students are taking demanding tests or when people find themselves in situations where they have to reason through a novel problem, giving their cognitive horsepower time to percolate can be beneficial.

PAUSING THE CHOKE

In the early 1980s, psychologist Michelene (Micki) Chi and her research team found that one big difference between students who succeeded and those who failed in difficult problem-solving situations was the time that they spent thinking about a problem at the outset—*before* they actually attempted to solve it. Jumping in at full speed can negatively affect your success. Chi was especially interested in problem solving in the sciences so she asked physics professors and some advanced Ph.D. students from the physics department at the University of Pittsburgh, where Chi was a senior research scientist, and several undergraduate students who had only completed one semester of the physics class, Mechanics, to solve several physics problems.[9] While everyone worked at completing the problems, Micki and her research team did something rather simple—they watched them.

As expected, the professors and Ph.D. students were better at solving the physics problems than were the undergraduates. Interestingly, however, the physics experts were not necessarily *faster* than the undergrads. Sure, once the professors and Ph.D.s got going on a problem, they were quicker to compute a solution. But Chi also found that the professors and Ph.D.s were slower than the undergraduates to *begin* to solve the problems. The experts paused before they ever put pencil to paper. They spent a few moments assessing the underlying structure of the problem and figuring out the best physics principle to use. The undergraduates, on the other hand, jumped right into problem solving, which often got them in trouble. By rushing to start the problem, the undergraduates got distracted by irrelevant problem details (such as whether there was a spring or a pulley mentioned in the problem description), which led them astray. The professors and grads deduced that they must find the major physics principle, for instance, "Force = Mass * Acceleration," as key to problem-solving success—and completely independent of whether there is a mechanism such as a spring or pulley in the problem.

Yes, the undergraduate students didn't have a lot of physics knowledge to begin with, so their problem solving may not have changed too much depending on whether they paused or not before they began to work out a problem. But for the physics professors and Ph.D.s who did have the knowledge and potential to excel on the problems, taking a step back before they began to work helped to ensure that they would not be led down the wrong solution path because of distracting information or silly mistakes. Pausing to assess the situation before starting to solve a difficult problem is one way to ensure success, especially if your first inclination is to look for the quickest and easiest way out.

If the VPs were to behave more like the physics professors and Ph.D.s than the undergraduates, this would help them find the best answer to the questions posed to them, especially when they were under pressure. In fact, even walking away from a problem for a few minutes can help folks find the most appropriate solution. This "incubation" period helps people to let go of their focus on irrelevant problem details and instead think in a new way or from an alternative perspective—producing an "aha" moment that can ultimately lead to success.

The legendary Greek philosopher, Archimedes, may have been the first person to demonstrate the power of taking a step back. Asked to determine whether a new crown made for King Hiero II was solid gold, Archimedes was under a lot of pressure to come up with the answer. He wouldn't have lost a college admission or a tournament title if he had failed, but there was a good chance he would have lost his life. Obviously Archimedes couldn't melt down the crown or break it open to determine its contents because that would destroy the crown. And because the crown was in the irregular shape of a laurel wreath, there was no object of a similar shape to which to compare it. Having puzzled over the problem of the crown, he did not find the answer until he stepped back from his task and stopped thinking about it altogether. As Archimedes was getting into the bath one day, he noticed that the level of water rose as he got in. He figured out that he could use the amount of water displaced by an object (either himself or the crown) to determine its volume. It was then an easy leap to divide the weight of the crown by its volume to come up with its density—which could help Archimedes determine whether the crown had dense gold or a less-dense silver inside. It was, indeed, the latter. According to tradition, Archimedes was so excited by his "aha" moment that he forgot to get dressed after he got out of the bath and ran through the streets naked yelling, "Eureka!"

Taking a step back rather than running full steam ahead when you have a task that requires a heavy dose of working-memory can be key to completing it successfully. Taking a step back can also be vital for carrying out problems that come up *after* you finish an immediately stressful task. The ability to perform difficult tasks declines over time—much in the same way that a muscle tires after exercise. In fact, glucose (which is a primary source of energy for the body's cells, including brain cells) becomes depleted when you continuously exert effort on a difficult thinking and reasoning task. If you don't take time to recoup your resources, your performance on whatever you do next can suffer.

> Take a step back instead of running full steam ahead when what you're doing requires a heavy dose of working-memory.

Glucose depletion may be especially problematic for those people highest in cognitive horsepower. Higher-working-memory individuals often tap a more extensive network of brain regions for performance than low-powered folks, so their brain cells need a lot of energy. This is because high-powered people tend to use the most cognitively demanding strategies to solve a problem. Recall that in the math test-taking experiment I told you about, above, it was those students with the most working-memory who tended to use difficult problem-solving strategies to find the mod math answers. So, pausing before you start to provide an answer or even pausing after you finish a task and go on to tackle the next project may make the difference between ultimate success and failure. Even though you might feel as if you don't have the luxury of catching your breath, going down the wrong solution path or operating with all "glucose cylinders on empty" is a worse option—especially for those people who have the most cognitive horsepower and the most potential to begin with.

GET USED TO IT!

During a coffee break at the Utah retreat, a woman from the audience approached me and asked, What else, besides taking a step back, can one do to help counteract the negative effects of stressful situations? I shared with her a story about my friend Raôul Oudejans, who is with the MOVE Research Institute at Vrije University in Amsterdam. Although Raôul is interested in all kinds of high-pressure situations, he has been spending a lot of time lately working with police officers to try to enhance their performance on the job.

While I'm sure the VPs have stressful lives, I don't think anyone would disagree that the types of pressures that cops face are of utmost urgency and that they often have to operate under the most pressure-filled situations at a high level of effectiveness. Raôul has found that training to shoot a handgun under stress helps to prevent skilled police officers from missing an important target when it counts.

In one study,[10] Raôul asked a group of police officers to practice shooting first at an opponent who was putting the pressure on by actually firing back—not with real bullets, but with colored soap cartridges. Raôul then asked these same police officers to take shots at cardboard targets (the kind you see cops practicing on in the movies). After the shooting practice, Raôul split his police officers into two groups. Half of the officers practiced firing at the live opponent and the other half only practiced shooting at the cardboard targets. Then everyone came back together and took some final shots—first at the live opponents and then at the stationary cutouts.

During the initial shooting practice, all of the officers missed more shots when firing at a live opponent compared with firing at the stationary cardboard targets. Not so surprising. This was true after training as well, but only for those officers whose practice had been limited to the cardboard cutouts. For those officers who practiced shooting at an opponent, after training they were just as good shots when aiming at the live individuals as they were when aiming at the stationary cutouts. The opportunity to "practice under the gun" of an opponent, so to speak, really helped to hone the police officers' shots for more real-life stressful shooting situations.

You might wonder if this type of "pressure training" is really effective, given that the stress simulated in training is not nearly as overwhelming as that of a real, high-stakes performance. Just think about the pressures a police officer faces when forced to shoot at someone who is firing back with real bullets rather than soap cartridges, or the pressure a professional soccer player feels when he is about to take a decisive penalty kick in the World Cup finals, or the pressure a high school senior feels as she sits down to take the SAT that will make or break her Ivy League dreams. Can you even begin to mimic the types of stressors that come into play in actual high-stakes situations? Yes, says, Raôul, because even practicing under *mild* levels of stress can prevent people from falling victim to the dreaded choke when *high* levels of stress come around.

> Even practicing under *mild* levels of stress can prevent you from choking when *high* levels of stress come around.

Regardless of whether you are shooting at an opponent in a police situation or shooting hoops in basketball, you can benefit from mild stress training. When people practice in a casual environment with nothing on the line and are then put under stress to perform well (let's say because a good chunk of money is now in play or their friends and colleagues will be watching their every move), they often choke under the pressure. But if people practice shooting a gun or shooting hoops or even problem solving on the fly with some mild stressors to begin with (say, a small amount of money for good performance or a few people watching a dress rehearsal), their performance doesn't suffer when the big pressures come around. Simulating low levels of stress helps prevent cracking under increased pressure, because people who practice this way learn to stay calm, cool, and collected in the face of whatever comes their way.

Indeed, these calm and cool qualities are the hallmark of professionals in many arenas and likely come from years of adapting to performing under pressure. A few years ago, a number of LPGA (Ladies Professional Golf Association) players visited the Brain Research Imaging Center at the University of Chicago.[11] The golfers were asked to imagine themselves taking a shot to a pin that was one hundred yards away, and while they were doing this, their brains were scanned using functional magnetic resonance imaging (fMRI) technology. fMRI assesses blood flow to specific areas of the brain, which can be used to gauge the amount of work that particular brain areas are doing while an athlete, say, takes a golf shot in her mind's eye.

When imagining their golf shots, the LPGA players activated a set of finely tuned brain areas involved in the planning and execution of actions. Golfers with only a few years of playing experience were also invited to have their brains scanned. When asked to imagine themselves taking the same hundred-yard shot, the less experienced golfers, in contrast to the professional players, activated a diffuse set of brain areas—including those involved in fear and anxiety. One hallmark of professionals seems to be a calm, cool mind, which researchers and coaches think comes from their experience practicing and performing under stress.

Practicing under the types of pressures that you are likely to face in an important game or tournament situation helps to ensure high levels of performance when it really counts. In my laboratory at the University of Chicago, we have shown that golfers who learned to putt in front of an audience were less anxious and putted better under stress than those who never practiced with others looking on. Thus that half hour of putting practice at the end of a round of golf may be more beneficial when it occurs while your friends look on than when you are alone—especially if you have to buy a beer for your buddies every time you miss a shot. This is also true for other activities such as public speaking or pitching a marketing campaign. Practicing answering on-the-spot questions before they actually have to face them in a real do-or-die situation may be just what the VPs and their team members need in order to get ready for the high-stakes meetings that their president is concerned about. If you are accustomed to operating under pressure to begin with, you will be less likely to choke, whatever you are doing.

Blue skies beckon the VPs to their afternoon activities, but as we are about to go our separate ways, several of them comment that they can already see how they are going to embrace some of the research I talked about in their business practices. Some are impressed by the idea of taking a step back when they hit a roadblock, others are going to get input from the employees who work under them before quoting delivery times to their clients, and all of the VPs seem to like the idea of getting their team ready for big presentations by practicing under pressure.

I have to head back to cloudy Chicago to teach a class of eighty undergraduates the next day and, as usual, I am banking on those few hours on the plane to finish up the lecture I will deliver. As we reach cruising altitude, however, I find myself thinking more about the talk I just gave than tomorrow's lecture. My morning in Sundance has made me realize that we psychologists have a lot to say to those outside our academic enclave. In fact, I had so much more I could have talked about. I should have emphasized to my audience that different types of stressful situations share commonalities, such as the fact that, as people's motivation to succeed increases, their likelihood of failing does, too, even though there are also differences in how people choke in the classroom versus on the playing field. I could have addressed how to practice to achieve the highest levels of performance and whether or not differences in innate ability carry implications for success. As I continue to think about everything that I didn't have time to convey, I decide that my lecture can wait, and I open up a blank page and begin to write this book. Now I can address all those questions.

CHAPTER TWO

TRAINING SUCCESS

When Carla decided to pursue a career as a concert pianist, she knew that she would be put in countless situations in which she would be called upon to display her musical ability on command. After all, a professional musician's livelihood is based on competing for seats in orchestras, giving concerts in front of discriminating audiences, and even doing on-the-spot sight readings of new musical scores for demanding composers. What Carla never imagined, however, was that, in addition to putting her music on show, she would one day be asked to do the same for the inner workings of her brain.

INSIDE THE HEAD: WHAT BRAIN IMAGING REVEALS

A lot of psychologists' work involves direct observation of human behavior, but in recent years we scientists have also come to use a wide array of tools to understand human performance. Brain-imaging technology, and in particular functional magnetic resonance imaging, or fMRI, allows us to see inside the brain when players are deciding which move to make next in chess, how best to execute a forehand in tennis, or how to finger a piano solo. We can see which brain areas are most involved in performance to get an idea of the ways in which great players harness brain power to succeed. This linking of brain to behavior has helped us to better understand the intricacies of success and failure—especially when the pressure is on.

In MRI, individuals lie with their heads and upper bodies inside a machine usually referred to as a scanner. An MRI machine is really just a large magnet—often fifty thousand times stronger than earth's magnetic field.[1] Nonetheless, surprisingly, a scanner's function is rather limited, but it can measure the magnetic properties of any object placed inside it. When we psychologists use MRI for our research, the object we're measuring in the scanner is the brain of the individual lying inside it.

MRI can look at anatomy because different types of tissue have different densities that it can detect and depict to form a picture of the brain. With a slightly modified type of MRI, called functional MRI, or fMRI, we can also infer what areas of the brain are working the hardest while people actually engage in thinking, reasoning, and problem-solving tasks.

Researchers had recruited Carla to the brain-imaging center in order to find out more about how musicians' brains are organized to support their skill and performance. But as Carla lay on her back and was moved via a motorized table so that her head and upper body were enveloped by the scanner's giant, hollow tube, all she knew was that she would be listening to a recording of herself playing Schumann's Allegro in B minor and that she was to play along with the recording using a makeshift keyboard set up in front of her.

What the researchers found was that the prefrontal cortex, the part of the brain where working-memory and conscious control are housed, does not perform the lion's share of the musician's work as some had believed it did. Rather, sensory and motor brain areas, where well-formed procedural memories reside, run the show during the performance of a well-practiced piece. Perhaps put best by Danish virtuoso pianist, Victor Borge, in a question to renowned pianist and conductor, Vladimir Ashkenazy, "Has it ever frightened you to play, and watch your fingers moving, and not know who it is that is making them move?"[2]

The prefrontal cortex of accomplished musicians isn't guiding their movements; it's their procedural memory, spinning outside their conscious awareness, that is doing so.

Information about how the brains of expert musicians function, and how beginners' brains operate differently, is precisely what researchers have used over the past several years to learn about how people in general come to be the best at what they do. Researchers have also learned something about what happens when pressure causes even the most skilled people to perform poorly.

Functional MRI (or fMRI) measures oxygen use of the active brain. Every activity activates cells in the brain called neurons, which create electrical pulses that result in tiny magnetic fields. Although these magnetic fields are far too small for fMRI to measure, fMRI can measure cells' oxygen use, for which neurons increase their demand as they become active and to which the body responds by increasing the blood flow to those active brain regions. Oxygenated blood has different magnetic properties than deoxygenated blood, which the MRI scanner can pick up. In other words, blood oxygenation varies according to the level of neural activity in a particular part of the brain and so this variation is thought to indicate brain activity.

You might expect blood oxygenation to decrease with brain activation as active brain areas use up more and more oxygen, but this is not the case. There is a momentary decrease in blood oxygenation immediately after neural activity increases, followed by a period where blood flow increases—not just to a level where oxygen demand is met, but to a level where these active brain areas are oversaturated. This blood flow peaks around six seconds and then falls back to a baseline state. So, fMRI doesn't directly measure neuron activity, but blood flow that occurs in response to this activity.

Functional MRI is an impressive technology, but it has its limits. For instance, neurons are very fast and blood flow is not. Neurons can send and receive signals in just a few milliseconds, which is good because important events in the world can happen in tens of milliseconds and neurons allow us to respond to these events just fine. However, the blood-flow response to a neuron firing takes around two seconds to get going, and around eighteen seconds to finish. Thus fMRI is good for capturing the location or particular area(s) in the brain that is highly active when people perform a task, but fMRI is not great for understanding the time course of a particular brain process.

Also, neurons are small and fMRI measures are big. There are approximately 10^{12} neurons (one thousand billion) in the brain, each of which is around one-hundredth of a millimeter. The smallest brain area (called a voxel) that fMRI can measure is usually a 3mm to 5mm three-dimensional volume. And usually when researchers make claims about brain activation, they are talking about several voxels clumped together. So fMRI data provides us with information about average activity across hundreds of thousands to millions of neurons—not very precise. When we use fMRI technology to infer that a particular area (or areas) of the brain is supporting the performance of some task, we rely on an assumption (a well-founded assumption, but an assumption nonetheless) that neighboring neurons are doing the same thing.

In this book I rely heavily on fMRI evidence to connect mental events to neural ones. Other techniques that we psychologists use to peer inside the head include electroencephalography (or EEG), which involves the subject wearing a cap full of electrodes to record the electrical activity along the scalp produced by groups of neurons firing within the brain. Because EEG directly measures neuron electrical activity (instead of the blood-flow response to this activity), it affords scientists more precise timing information about a neural event than fMRI does. Yet because neural activity is collected from millions of neurons at once, information about which brain areas are carrying out the neural event is not very precise.

Before MRI technology was widely available, scientists largely relied on another brain-imaging method known as positron emission tomography (or PET), which requires that radioactive tracers be injected into the subjects' bloodstream and measured while they do different perceptual, cognitive, and motor tasks. Just as with fMRI, with PET we assume that the brain areas that are working harder are recruiting more oxygen and other important nutrients such as glucose. We also assume that when a particular area of the brain needs more oxygen and glucose, the body's vascular system will shunt more blood to that part of the brain to satisfy its needs. fMRI measures the oxygen content of this blood flow while PET tracks the blood flow itself. Because PET is quite invasive (with its injections of radioactive elements), however, many researchers today opt for the noninvasive fMRI technique over PET.

fMRI images are certainly fascinating, but they should never be credited—at least in this early stage of the technology—for the ability to decipher with 100 percent precision what is going on inside the head. A San Diego–based company, No Lie MRI, claims that MRI can be used as a "direct measure of truth verification and lie detection." Others have suggested that technology like fMRI is the equivalent of mind reading. Although such claims are tantalizing, the fact of the matter is that a brain scan is just a brain scan. It provides information about neural activity, but doesn't on its own explain the complexity of human behavior, including moods, motivation, intentions, decision making, and anxiety. Brain data should ideally be seen as just one piece of the puzzle of human performance.

Maybe people are so fascinated with fMRI and other neuroimaging techniques because we don't yet know their true potential. Just as we hope to find the next star when a new athlete emerges from a stunning upset, we hope for limitless information from each technological breakthrough. Think of the story of American tennis player Melanie Oudin. Melanie was virtually unheard-of on the professional sports scene until she caused a series of surprise upsets against top-ranked players including Maria Sharapova at the 2009 U.S. Open. These upsets landed Melanie in the quarterfinals and, all of a sudden, the blond, blue-eyed teenager from Marietta, Georgia, had become a household name. Speculations that she was the next tennis superstar abounded.

Some argued that Melanie Oudin's popularity stemmed from her perky attitude, the fact that she had the word *believe* written on her sneakers, or her habit of coming from behind to win with a vengeance. But I suspect people have rallied behind Melanie because no one yet knows what she can do. A year before the 2009 Open, Melanie was still playing in the girls' tournaments. She had nothing to lose at her first Grand Slam advanced-round appearance and everything to gain. Melanie Oudin has yet to show what she is really made of and that has everyone cheering on her side.

But until Oudin is winning Grand Slam titles, we should be careful about the significance we attribute to her U.S. Open performance. And we should be very careful in attributing too much significance to the data depicted in brain imaging. As psychologists David McCabe and Alan Castel recently discovered, sometimes people put too much meaning into pictures of the brain.[3] These researchers asked undergraduates at the University of California at Los Angeles and Colorado State University to rate the scientific quality of psychology articles that were accompanied either by figures depicting areas of the brain that were active when people performed different cognitive tasks or by bar graphs that conveyed the same information as revealed in the brain images.

Although the undergraduates didn't know it, the articles they read were fictitious. The researchers had made them up to test the readers' ability to spot disparity between the articles' claims and the data the articles presented. For example, one article, titled, "Watching TV is related to math ability," concluded that, because both watching television and doing arithmetic problems lead to activation in the temporal lobe of the brain (the temporal lobe is generally involved in memory and visual attention), watching television could be used to improve math skills. Although I am sure that the undergraduates in the study would have been happy to find out that TV was so beneficial, just because the same area of the brain is active during two different tasks doesn't mean that doing one activity (here, watching TV) will improve performance on a different type of task (for example, math problem solving).

Sure enough, the articles that the students rated higher in scientific quality presented brain images alongside the text. Articles accompanied only by bar graphs received lower ratings. Attributing greater significance to brain scans than they warrant also occurs in real science, not just made-up work. BBC news service wires that summarize psychology work are rated by readers as more credible when the wires are accompanied by flashy pictures of the brain, compared to reports that do not contain such illustrations. A picture is thought to be worth a thousand words, but as a discriminating reader, you have to make sure those words describing the image are accurate.

One reason that brain images may be so influential is that they provide a physical basis for thinking about abstract notions such as the mind. In a sense, brain pictures help to make more tangible abstract constructs such as memory and attention and to simplify the obviously complex processes of thinking and reasoning. Yet it is really important to understand that these images simply illustrate data the way a bar graph or pie chart does. The interpretation of these brain images can be just as subjective as the conclusions drawn from other types of scientific data.

Nonetheless, despite its limitations, brain imaging technology like fMRI does supply useful information, including what happens in the brain when people fail to perform at their best under stress. By identifying the brain regions or brain networks most affected by high-pressure situations, you can gain insight into how choking under pressure occurs and what you can do to prevent it. Brain-imaging work has also provided new insights into how talents and skills develop. As we will see in the pages to come, practice can actually change the physical wiring of the brain to support exceptional performance.

Practice can actually change the physical wiring of the brain to support exceptional performance.

PICKING IT UP

Dan had always been a stellar athlete. From the start of elementary school through his college years, every time Dan tried his hand (or foot) at a new sport he quickly became the best player on the court or field. Dan didn't excel only in athletic play; his sport prowess extended off the playing field as well. He could tell you the history behind most sports teams or talk your ear off for hours about tactics. This knowledge came in handy during trivia games or when Dan's friends were arguing about which major-league pitcher—Roger Clemens or Nolan Ryan—held the record for the most career strikeouts (it's the latter). Perhaps most impressive about Dan's sports ability was that, at least from the outside, he didn't seem to have to work very hard to excel. Dan was strong, quick, and had good anticipation skills. He was always in the right place at the right time, whether it was on the court to make the game-winning basket or on the soccer field to halt an opponent's run for the goal. How did Dan do it? And, more generally, how do superstar athletes—ranging from golfer Tiger Woods to tennis player Steffi Graf—manage to rise to the top of their games?

Sports fans, media commentators, and even athletes themselves often wonder whether innate athletic gifts or training creates superstars. Throughout the summer of 2008, for example, many wondered how forty-one-year-old Dara Torres had managed to come back to her sport and win a medal at the Beijing Olympic Games after seven years off from competition and a pregnancy to boot. It was Torres's fifth Olympics, and she was the first American swimmer to achieve this feat. Could anyone who invested the time and monetary resources that Dara had invested also achieve swimming greatness, or were there other, rare qualities needed for Dara's level of success?

The origins of ability and skill have been passionately debated in business, music, education, and the arts, but nowhere more heatedly than sports. Why are some able to achieve athletic feats that others cannot? Are athletic superstars born or made? Uncovering how athletes cultivate extraordinary abilities in their sports can certainly provide clues for developing the next generation of superstars and even help weekend athletes perfect their golf game or overhand shot. Knowing why the best

> Knowing why the best are the best can also tell us something about who will *thrive* and who will *dive* when the stakes are high.

are the best can also tell us something about who will *thrive* and who will *dive* when the stakes are high—and what you can do when you find yourself more so on the "dive" than "thrive" end of things.

Dan attended college at a large midwestern university where he was a soccer star. Not only was Dan one of the leading goal scorers on the team all four years of his college career, but he made the all-American team, the best of the best in the United States, three years running. Dan had played several different sports growing up, ranging from baseball to tennis, but soccer had always been his first love and the sport in which he had truly excelled. Of course, in the United States, soccer doesn't carry the same passionate fan base as in other parts of the world, nor does excelling in soccer have the same cachet that it does everywhere else. But Dan didn't care. He loved the game, never missed an opportunity to play, and tried to learn as much about the sport as he could.

Dan was excited to discover a newspaper article about Dr. Werner Helsen, a European sports psychologist who used to play soccer and now makes his living uncovering the secrets of soccer success at the Katholieke University in Leuven, Belgium. Helsen believes that if you can figure out what differentiates "sport greats" from "weekend warriors," then you can use this knowledge to propel folks to the top on the playing field and hopefully keep them there—even under the most pressure-filled circumstances. Helsen conducted one of the first comprehensive sport science studies designed to answer the question, Are soccer stars born or made? In other words, do professional players come into the world poised for soccer greatness? Or do hard work, practice, and experience enable them to become stars of the world's most popular game?

To get at these questions, Werner and his collaborator, Janet Starkes, examined whether professional soccer players' basic visual and motor abilities are superior to recreational players' abilities or whether elite players *only* surpass recreational players when they perform skills specific to the soccer game.[4] After all, soccer players don't spend a lot of time in laboratories doing basic motor tasks like reacting to lights flashing on a computer screen. They practice soccer-specific skills like passing, shooting, and reacting to their opponents' moves. Showing that professionals outshine recreational players in basic athleticism would suggest that being "naturally athletic" is important for achieving elite status. On the other hand, if professionals *only* outshine recreational players when they are performing soccer skills that they practice all the time, then this would suggest something quite different. Instead of athletic prowess being endowed from the heavens, it may be that stars are primarily made through the thousands and thousands of hours they spend on the field.

The researchers asked both professional and recreational soccer players to perform a variety of tasks in the laboratory that tapped either basic perceptual and motor coordination abilities or more soccer-specific skills. For instance, players were asked to respond as quickly as they could (by pressing a button positioned in front of them) when a green light was flashed on a computer screen. In another basic task, players were asked to visually track an object moving across a screen (such as a chevron, that is, a < or > sign). The players' goal was to keep the object in view so they could report, as fast as possible when asked to do so, whether the chevron was pointing to the left or to the right. I think it is pretty safe to say that soccer players don't practice these tasks on a daily basis.

Interestingly, Helsen and Starkes didn't find any meaningful differences between the professional and recreational players on these tests of basic perceptual and motor abilities. The professional players were no better than the recreational folks at reacting to lights or tracking meaningless objects. However, things changed when the players were asked to perform visual and motor tasks that were related to soccer.

Players were put in a virtual reality soccer simulator and given a real ball. They had to react to oncoming opponents on a life-size screen in front of them and decide what move to make—to shoot, pass, or dribble away. They still had to respond quickly and track objects across a screen (just as they had tracked the chevron in the basic ability tasks), but now they were tracking players as they might move across the field in an actual game.

In this soccer context, the professional players picked up on important visual information that less experienced players missed—such as the presence of an open player on the other side of the virtual field in front of them. The professionals were also able to react more quickly and make better tactical decisions when watching a play unfold on the screen in front of them compared to their recreational counterparts.

Everyone, regardless of his soccer prowess, had the same athletic abilities when tested in a general context, but they differed in their abilities to track objects and react quickly in a situation more akin to a soccer game. So it seemed like a good bet that the soccer players whom Helsen and Starkes tested were experts primarily because they had learned skills specific to their game.

It's easy to find anecdotal support for the importance of practice. Just think about basketball great Michael Jordan. The way he flew through the air to dunk a ball on the court gave the impression that he was endowed with extraordinary athletic abilities that he could use generally in other sports. But Jordan's failed stint as a professional baseball player told a different story. Jordan spent 1994, the year after his first retirement from professional basketball, pursuing a childhood dream of being a professional baseball player. Jordan landed on a Chicago White Sox farm team, the Birmingham Barons AA club, and had a nonstellar baseball season, ending with a mediocre batting average of .202. Jordan couldn't hit a curveball with an ironing board. If Jordan's biologically endowed motor abilities drove him to success on the hoops court, why would they fail him on the baseball diamond? After all, quick reactions, agility, coordination, and power are needed to excel in both sports. But practice seems to have honed Jordan's basketball skills and a lack of practice seems to have limited his baseball success.

Of course, the debate over whether stars are "born" or "made" has been going on forever or at least since the time of the ancient Greek philosophers. Plato argued that we come into this world with biologically endowed abilities and skills and that our highest levels of success are predetermined by the heavens. On the other side of the debate was Aristotle, who just happened to be Plato's student, and who adamantly believed that success was gained through learning and training. Several modern-day researchers, including Werner Helsen and Janet Starkes, take Aristotle's side, but not everyone does. Australia, for example, in its quest for international sports greatness, seems to be betting on a combination of Plato and Aristotle in their approach to intensively training certain natural talents.

SPORTS DOWN UNDER

Australia treats sports as a science and uses all the scientific tools at its disposal to develop exceptional performers. The island nation's tools range from nutritional analysis of the foods that lead to optimal endurance in distance swimming to psychological analysis of the best training techniques to ensure rapid decisions in team handball. Although virtually unheard-of in the United States, team handball is of major interest on the international sports scene. Much of Australia's sport science work is done at the Australian Institute of Sport (AIS), in the national capital, Canberra. The AIS is one of the development centers for Australia's elite athletes, and also houses a team of sport scientists whose goal it is to unlock the psychology, biology, and physiology behind world-class performance.

In the early summer of 2005, I was invited to the AIS, along with several other sports scientists trying to uncover the secrets behind exceptional performance, to spend a week with some of Australia's national and Olympic coaches. The plan was that we would trade insights about how to develop the best athletes and, most importantly, how to keep them at the top once they had arrived. While at AIS, I learned about an aggressive talent identification program that the Australians have adopted in several sports in an attempt to increase their medal count on the world sports scene. One program I found especially interesting was under way in the winter Olympic sport of skeleton.

On a bobsled track, skeleton competitors must push their sleds quickly and then dive onto the sled headfirst and pilot it through

rough curves at very high speeds facedown. The Australians have taken athletes with impressive sprint-running ability and transferred them into the sport. Their idea is simple and, at first glance, really seems to align with Plato's view that athletes are born rather than made. Basically, because explosive speed is so important for skeleton success, if you are fast you need little training. The Australians have had some remarkable success with this approach. For instance, Michelle Steele, after roughly four months in the sport, placed sixth at the 2005 World Cup held in Calgary.

The idea that sports talent is innate is seductive. But a closer look at cases like Michelle Steele's reveals that practice plays a bigger role than you might initially think. The Australian skeleton competitor put in serious practice time once she committed to her new sport. And before she ever thought about whisking down an icy skeleton track, she had been developing her powerful sprinting ability by logging long training hours as a nationally competitive gymnast.

Of course, there are differences between individuals: kids come in bigger and smaller sizes and some naturally do have more working-memory than others. But despite innate differences, our eventual level of success is markedly affected by training and practice. Certainly, if *everyone* gets the same type of training or educational input and *everyone* improves by the same amount, then any individual differences in size, speed, or cognitive ability at the beginning of practice will still be there at the end. But this is usually not the case. Without the intensive training that the Australians provided for Michelle Steele—both in the gym and then on the track—she would never have become a world-class skeleton contender (placing thirteenth at the 2006 Winter Olympics; second in Nagano's 2007 World Cup; sixth in World Championships in 2007). Without the hundreds of thousands of dollars Dara Torres shelled out for coaches, trainers, doctors, and practice time, she would never have made her latest Olympic comeback.

Despite innate differences, our eventual level of success is markedly affected by training and practice.

Unfortunately, Michelle missed making the cut for the 2010 Australian Olympic skeleton team. She was outpaced just weeks before the Winter Games by Australian teammates Emma Lincoln-Smith and Melissa Hoar, who finished tenth and twelfth, respectively in Vancouver. Interestingly, like Michelle, neither Emma nor Melissa started her athletic career sliding. Instead, they were both plucked from surf lifesaving, a competitive ocean sport involving beach sprint and relay events and demanding the same type of explosive speed needed to dominate on the skeleton track.

Perhaps because of the growing awareness of the importance of intense practice, it's not unusual these days for a kid to spend his or her summer at a series of athletic camps specializing in one sport, say, soccer or lacrosse. Some parents even shell out upwards of fifty thousand dollars a year to send their kids to sport-specific academies like IMG in Bradenton, Florida. At IMG, kids spend just as much time in a classroom as they do practicing their sport with teammates, as well as in professional coaching, video analysis, and intense physical conditioning.

In some instances this type of intense training and early specialization works quite well. Tennis player Maria Sharapova, who at age seventeen won Wimbledon, trained at Nick Bollettieri's Tennis Academy at IMG. But despite the fact that practice is central to winning in sports, early training and specialization may not always be the way to go. Other factors also play a large role in success.

BIRTH DATE AND BIRTHPLACE MATTER

Couples who are thinking about having a child consider many factors related to timing. How will pregnancy and having a newborn affect their work and travel schedules, and their lifestyle in general? Are they financially and—most important—psychologically ready for this life-changing event? Of course, most couples don't consider whether having a child in the early or late seasons of the year might affect their youngster's ability to succeed in school or on the playing field. But it does.

Dan's birthday is August 1. This is important because the birth date cutoff for participation in most sports in the town where Dan grew up is July 31. This meant that Dan always missed the cutoff to play on the older team and ended up on the younger team—competing against kids who were often close to a year younger than he was.

> The older a child is relative to his peers in soccer, the greater the probability this child has of eventually becoming an elite player.

Birth date cutoff is quite typical in sports. In fact, most teams specify birth dates to ensure that there are not large age differences within a single unit. Nonetheless, children born right after a particular birth date cutoff can be a lot older than late-born children in the same cohort and this age difference has consequences for success in sports. Researchers including Werner Helsen, the sport psychologist at Katholieke University in Belgium, have shown that. Werner calls this phenomenon the *relative age effect*.[5]

Relatively older children are often developmentally advanced for their peer group. They are more coordinated and athletic, which in turn leads to earlier identification and selection for sport training, which can propel them to the next skill level. Exposing kids at an early age to competitive environments that they must thrive in later on also helps them adapt to performing when the stakes are high. As we saw in chapter 1, training under pressure helps ensure success when it counts the most. Being bigger than your teammates because you are almost a full year older than many of them makes a difference as well. Coaches notice and this often leads to that extra bit of playing time, extra confidence playing in important pressure-filled situations, and ultimately extra skill success.

There are similar age effects in the classroom. Children's grasp of a concept called conservation of number, which in a nutshell is a child's understanding of the fact that the number of objects in a group remains the same despite changes that are irrelevant to quantity—

such as simply moving the objects around, shows a relative age effect.[6] Not only do older kids in a grade have more time to learn about number conservation outside the classroom, but cognitive abilities like attention and memory (abilities that are necessary to support one's understanding of number) increase with age. Relatively older children are at an advantage because they get more informal learning outside the classroom and their brains are developmentally more advanced, so what they learn in the classroom actually sticks. If these older kids get earmarked as "smarter" or more "intelligent," their relative age could help them stand out from their peers.

Interestingly, it is not just a child's age that can lead to an advantage. Where a child is born can also be a key to his or her eventual success. A group of sport scientists in Canada recently discovered this *birthplace effect* while poring over the statistics of athletes in the National Hockey League (NHL), National Basketball Association (NBA), Major League Baseball (MLB), and professional golf (PGA).[7] Using the data from more than two thousand male professional athletes in the United States and Canada, the scientists discovered that the percent of athletes who came from cities of fewer than a half-million people was higher than what would be expected by chance alone. In contrast, the percent of professional athletes who came from cities with more than a half-million inhabitants was a good deal lower than what you would expect by chance. While nearly 52 percent of the United States population resides in cities with more than a half-million people, such cities only produce about 13 percent of the players in the NHL, 29 percent of the players in the NBA, 15 percent of the players in MLB, and 13 percent of players in the PGA.

While nearly 52 percent of the United States population resides in cities with more than a half-million people, such cities only produce about 13 percent of the players in the NHL, 29 percent of the players in the NBA, 15 percent of the players in MLB, and 13 percent of players in the PGA.

Smaller cities offer more opportunities for unstructured play, which leads to longer hours of practice and involvement in sports at a young age. A kid in a smaller city can spend hours in the park by himself or kicking a ball against her parents' garage—something that is hard to do in the middle of a big city. Practice comes in many forms.

Perhaps because there is less competition to make any one team, children in smaller cities also get the opportunity to sample many different sports. Trying out a variety of activities lowers the likelihood of burnout in one sport and increases children's feelings of confidence because they get to see the results of their hard work in different settings. Playing different sports also lessens the occurrence of sports-related injuries that may end an athletic career. For instance, it's common today for a ten-year-old baseball pitcher to need tendon replacement surgeries for an injured elbow—these were previously restricted to college and major-league pitchers. This is the type of injury that, many sports medicine doctors argue, is the direct result of arm overuse and sport specialization at too young an age.

Findings like the birthplace effect suggest that we need to rethink the growing trend for kids to receive year-round training in one sport early on. Instead, less sport-specific training and more diverse recreational play seem to be preferable for developing athletic ability and expertise. Of course, this doesn't mean limiting practice overall. It does mean, however, that there are better ways than early sport specialization to hone one's craft. For Australian champion Steele, practicing her sprinting ability early on in the gym may have been the ticket to her success on the skeleton track. Who knows what would have happened if she had specialized in skeleton too young—she may have been injured, gotten burned-out, or both.

> Less sport-specific training and more diverse recreational play may be best for developing athletic ability and expertise.

PRACTICE, PRACTICE, PRACTICE

Despite the fact that a child's birthplace or birth date might affect his or her eventual success, the main factors that separate extraordinary performers from ordinary ones are the time and effort they put into developing skills they will eventually need to excel at their craft. Even though it might seem that some people don't train very hard to succeed, they are likely logging thousands of practice hours. This training may not always be in a formal team setting, and as we just saw above, that is probably a good thing. But that doesn't mean that practice isn't important. In fact, practice is key in all sorts of activities—even activities that, at first glance, might seem to be largely driven by innately endowed abilities.

Think about skilled players of a game like chess. Recreational players in the park who play speed chess seem to recall an unlimited number of possible moves all within the window of a few seconds. Another type of competitor, the professional chess master, may take longer to play any given move than the speedy park players, but as a master ponders his next move, he too is going through many possible game scenarios in his head. Certainly you might think that chess masters who operate at the highest levels of one of the world's most popular games, or even the skilled park players, have extraordinary memories that help them. Yet, just like on the skeleton track or soccer field, practice (rather than any sort of innate or endowed ability) largely drives greatness in chess.

In the 1960s, a psychologist who also happened to be a Dutch chess master—Adrianus Dingeman de Groot—ran a series of experiments that showed that practice was a major determinant of chess success.[8] Interestingly, this discovery didn't come easily to him. De Groot was initially unable to find obvious differences between chess masters and less skilled players in important aspects of the chess game. Masters and moderately skilled players, for example, didn't differ in the number of moves they considered during play or the persistence in which they searched their memories for possible next steps. But then

De Groot ran an experiment that revealed something about what it takes to be a chess master.

De Groot showed chess masters and weaker players a chessboard on which the pieces were configured as they might be in the middle of an actual game. After about five seconds, De Groot removed the pieces and asked the players to reconstruct the board—exactly as they had just seen it—from memory. Even though they had seen the board for only a few seconds, chess masters were able reconstruct it almost perfectly, but players below the master level had a lot of trouble.

De Groot then showed masters and weaker players a second board that was not representative of a real game, and on which all the chess pieces were randomly mixed together. The chess masters' ability to reconstruct the random boards was just as bad as that of the less skilled players. The masters *only* showed extraordinary memory for chess situations that they might see in an actual game. In other words, chess masters don't have superhuman memories that support their reasoning, strategizing, and move selection in chess. Instead they seem to have learned specific tricks that help them to remember and reason through realistic games.

In 1973, two psychologists from Carnegie Mellon University, William Chase and Herbert Simon, figured out exactly what these "tricks" were. That is, they learned how chess masters are able to reconstruct an entire chessboard from memory when the pieces are positioned in a situation that might actually occur in a game *and* why their memory fails them when the pieces are randomly arranged. These researchers took seriously the players whom they interviewed who said, "I just see the right moves," and they looked at how masters and less skilled players visually examine the chessboard.

The researchers invited a chess master and a weaker player to visit their psychology laboratory at Carnegie Mellon.[9] When the players arrived, they asked them to sit at a table where two chessboards were placed in front of them in plain view. To the left was a board depicting a realistic midgame chess scenario and to the right, a blank board. The researchers asked the players to reconstruct the chess pieces that were on the left-hand board on the blank board on the right. While the players did this, the researchers watched them in the hopes of inferring—from how the players glanced from one board to another—what the players actually saw.

Watching the players watch the boards proved surprisingly informative. The researchers found that the better the chess player, the fewer glances he needed from one board to the other to reconstruct the pieces. With each glance, the chess master seemed to capture several different chess pieces as a group. Because one group of pieces is easier to remember than nine individual pieces, for example, the chess master was able to reconstruct the full board in less time and with fewer glances than the weaker player. The chess master was seeing individual chess pieces organized together in some sort of meaningful way—for example, a certain attack sequence designed to capture an opponent's rook. By doing this, he had fewer separate pieces of information to hold in memory, which made it easier to hold more. Because the chess master worked by finding meaning in the pieces, when the pieces were presented randomly and couldn't be grouped into game-specific patterns, the master's memory looked like that of the less-skilled player.

The chess masters' enormous amount of practice, not simply their extraordinary memories, allows them to see meaningful patterns in the board that less skilled players can't see. These patterns can help masters think ahead ten moves whereas a less skilled player can only think three moves out. Perceiving patterns may even help the chess master anticipate his opponent's next move before the opponent himself realizes what he's going to play.

Chess masters aren't alone in using tricks to circumvent normal memory limitations. Waiters who can remember several dinner orders without writing a single one down also use memory aids. Take the famous case of a waiter who went by the initials JC and could remember up to twenty dinner orders in a row without writing a single one down.[10] Psychologists studying memory ability were quite interested in him and found that JC's extraordinary memory wasn't an innate gift, but was due to some tricks he had learned to hold lots of information in his mind at once. Let's say JC was taking a dinner order from a table of four—which was easy for him. Rather than trying to remember each dinner order separately, JC would organize the orders at a table into meaningful groups so that he didn't have to remember every little detail on its own. For instance, if everyone at the table ordered a salad—one person asked for ranch dressing, another Italian, a third Thousand Island, and a fourth vinaigrette—JC would remember RITV. JC developed a mnemonic for the dressing that placed less demand on his memory than having to remember each dressing separately. Just as seeing nine separate pieces on a chessboard as part of one meaningful attack sequence allows a chess master to turn many pieces of information into one, JC created patterns for dinner orders that helped him to pare down a lot of information bits into manageable groups.

This type of mnemonic strategy can aid anyone trying to remember information for a big test or presentation. Finding meaningful ways to group separate pieces of information into smaller bundles can take the burden off working-memory and help you remember more. Bundling information can also be advantageous for performing under pressure, say in an important test or a do-or-die pitch to a client. Worries and self-doubt flood the brain when the pressure is on and tax the very memory components that we use for keeping track of lots of different pieces of information. Combining what you need to remember into meaningful wholes helps to ensure that some pieces don't get lost when it counts the most.

> Group pieces of information into bundles to help you remember them.

PRACTICING THE BRAIN

Practice can help you train your basic perceptual, cognitive, and motor abilities to your advantage. It can aid memory by helping you find relations between pieces of information that might seem very disparate. Practice can also change how your brain is wired to support exceptional performance.

Take cabdrivers, for example. They learn every street in a city because they practice. Big-city cabdrivers spend several years memorizing various ways to navigate their crowded metropolitan area before they are allowed to set foot in their own cab. Scientists have shown that this route-finding practice changes these cabdrivers' brains.

The hippocampus, which is important for navigating and recalling complex routes, is enlarged in London cabbies compared to nondrivers.[11] Even more telling about the role of practice in changing the brain is that the size of cabdrivers' hippocampus varies with years spent behind the taxi wheel. The longer a London driver has been on the streets, the larger the part of the hippocampus involved in successfully finding the correct city route.

Brain training produces similar effects in the art of juggling.[12] Several months of juggling practice increases gray matter (where the cell bodies of neuron are housed), which generally means greater communication among brain cells, in parts of the brain involved in understanding motion. Interestingly, when folks stopped their intensive juggling practice, the motion-understanding brain areas, which had changed in density, changed back to become less dense. Just as lifting weights helps to develop your bicep muscles, practice shapes your brain. However, these practice-related changes often only stick around if, like working the biceps, you continue to work your brain.

Just as lifting weights helps to develop your bicep muscles, practice shapes your brain.

Practice-induced brain changes occur in the musical world, too. Many musical instruments require fine-tuned coordination of both hands. Because each half of the brain (or hemisphere) largely controls the opposite side of the body, the two hemispheres need to talk to each other in order to coordinate hand movements. This talking is primarily done though a wide bundle of nerve cell fiber tracts that connect the two halves of the brain together: the corpus callosum.

Interestingly, musicians who began their training early in life have a larger corpus callosum than those who started training later. Musical training, and especially early musical practice, can enhance the interaction between the two hemispheres of the brain. Early musical training is also related to the attainment of absolute or perfect pitch—the ability to reproduce and recognize musical notes without any sort of external referent.[13]

Musical training, and especially early musical practice, can enhance the interaction between the two hemispheres of the brain.

Why might this be the case? One idea is that the early learning of instruments such as the violin or piano is less dependent on the prefrontal cortex, which becomes more involved when these same instruments are learned later in life. Because the prefrontal cortex develops with age (this brain area isn't thought to reach full maturity until well into early adulthood), when people learn skills early, other brain areas such as sensory and motor cortex take over. Learning earlier helps with the acquisition of skills—for example, absolute pitch—that are best performed with a heavy dose of input from sensory and motor brain areas.

These same learning mechanisms are in place in the acquisition of language accents, too.[14] It's no secret that we tend to have better accents for languages that we learned when we were young children. Scientists think this happens in part because the words we learn as kids are more closely linked to sensory and motor brain areas than words learned as adults. Because these sensory and motor areas are involved in processing the sounds of the words and speaking the words, reproducing correct words and their accents is easier when these brain areas do a lot of the work.

In sports, my collaborator Arturo Hernandez and I have shown that the age at which training commences in golf plays a role in how the brain supports putting. We found that skilled golfers who started playing golf *after* age ten rely more on working-memory during the execution of a simple putt than those who started earlier. Even though we handpicked all of our golfers so that all were equally good (single-

> The later golfers learn, the more vulnerable they are to choking under pressure.

digit PGA handicaps), the age at which golfers begin learning and practicing affects how their brains help them play.

We also think that, the later golfers learn, the more vulnerable they are to choking under pressure. As we will see in the chapters to come, athletes under pressure sometimes try to control their performance in a way that disrupts it. This control, which is often referred to as "paralysis by analysis," stems from an overactive prefrontal cortex. One way to circumvent this type of paralysis is to employ learning techniques that minimize reliance on working-memory to begin with. When you start playing earlier in life, your prefrontal cortex may not be as likely to get overinvolved when you're under pressure. Those who take up golf in the first several years of life may be in the best position for success under stress. Of course, as a way to circumvent some of the problems that arise due to early specialization, it's probably better to begin this early golf training in parallel with other sport training, too.

> Athletes' tendency to overthink their performance is one big predictor of whether they will choke in important games or matches.

Dr. Richard Masters and his colleagues who run the Institute of Human Performance at the University of Hong Kong believe that an athlete's tendency to overthink their performance is one big predictor of whether they will choke in important games or matches.

Masters asks athletes to rate questions such as those in the box below from *strongly disagree* to *strongly agree.* He has demonstrated that how people respond to these questions predicts their propensity for poor performance under pressure.[15]

For instance, Masters found that university squash and tennis players who were rated by their coaches as likely to "choke under pressure" were more likely to agree with the sentiment of the above questions than those whose coaches viewed them as solid "go-to" players under

1. I rarely forget the times when my movements have failed me, however slight the failure.
2. I'm always trying to figure out why my actions failed.
3. I reflect about my movement a lot.
4. I am always trying to think about my movements when I carry them out.
5. I'm self-conscious about the way I look when I am moving.
6. I sometimes have the feeling that I'm watching myself move.
7. I'm aware of the way my mind and body works when I am carrying out a movement.
8. I'm concerned about my style of moving.
9. If I see my reflection in a shop window, I will examine my movements.
10. I am concerned about what people think about me when I am moving.

stress. Recently, Masters and his colleagues have shown that people with Parkinson's disease agree more with these questions than those who don't have this neurological degenerative disease. The longer people have had Parkinson's, the more they endorse the above statements. Parkinson's disease is typified by difficulty in movement initiation and execution, which often causes patients to consciously monitor their actions. In sports, this type of performance monitoring may be one cause of severely disrupted movements, such as the "yips" in golf. We will come back to this idea in chapter 7.[16]

WHERE ARE WE NOW?

Dan, our soccer star, spent a lot of time on the field honing his skills so that he could use his speediness to his advantage. His birth date—being relatively older among his peer group—gave him advantages because he was ahead of the other kids in his coordination and athletic ability, which led to earlier selection for soccer training opportunities and more experience adapting to high-stakes competition. Dan also had a lot of opportunity to engage in informal play, likely limiting sport-specific injuries, burnout, or both.

All these practice-related factors have an important role in shaping a child's success. Understanding how one gets to be the best—on the court, on the stage, in the boardroom, or in the classroom—is interesting in its own right, but it is also important for revealing how and why performance breaks down under stress. Before we explore performance failure (and what can be done to turn around disappointing outcomes) in more detail, however, we have to cover a few more topics about acquiring high levels of skill.

This chapter emphasizes the role of practice, but we can all agree that people differ in their innate core cognitive and motor capacities. For instance, Finnish cross-country skier and three-time Olympic champion Eero Mäntyranta has a genetic mutation that increases his hemoglobin concentration and, as a result, promotes enhanced oxygen supply to the brain and muscles.[17] Certainly the Finnish skier's genetic makeup has helped him perform at the top of a sport in which endurance is of utmost importance.

What has science discovered about these genetic differences and how they influence advancement in sports, education, and work performance? Even if we can train away naturally occurring variation with practice—is it always better to do so? In the next few chapters we tackle these issues.

CHAPTER THREE

LESS CAN BE MORE

WHY FLEXING YOUR PREFRONTAL CORTEX IS NOT *ALWAYS* BENEFICIAL

Sara grew up in the foothills of Oakland, California, in a spacious house that her family absolutely adored on a quiet cul-de-sac, with an amazing view of San Francisco Bay. Even though Sara's family loved their home, the summer before she entered seventh grade they moved a few miles down the road to the city of Piedmont. The move was based on one simple factor: the Piedmont public schools.

Unless you live in the San Francisco Bay area, you probably haven't heard of Piedmont—or its school system. This is because Piedmont is a small bedroom community of about eleven thousand residents that is completely surrounded by the city of Oakland. Although Oakland and Piedmont are geographically intertwined, in many ways they couldn't be more different. For one thing, the median price of a home in Piedmont is roughly three times that of a home in Oakland. The main reason for this price difference is each city's school system. Piedmont schools rank near the top of all California public schools, whereas Oakland schools consistently rank closer to the bottom. When people buy a home in Piedmont, they don't just pay for the place in which they live; they also pay for the privilege of sending their kids to the outstanding schools.

Sara's parents obviously viewed the school environment as a major determinant of her academic achievement. As we saw in chapter 2, the merits of training and practice are well founded. Nonetheless, even though a good education has undeniable benefits, you have to wonder just how *much* environment influences success. Like Sara's parents, some people spend thousands of dollars to move to affluent neighborhoods so that their children can attend good public schools; others pay high tuition for top-notch private schools. Not everyone can afford to live in the "right" school district or to pony up private school money, however, and as a result, different kids grow up getting educations of varying quality.

If educational experience were the sole predictor of academic accomplishment, we would expect that Sara would be off to a top Ivy League university once she graduated from Piedmont, while the friends she left behind in Oakland would be attending far less prestigious schools. This is not what happened. After high school, Sara went to Chico State University, which is less academically rigorous than institutions such as Stanford and UC Berkeley, where several of her Oakland pals matriculated.

So, environmental influences don't explain everything. Just as some people are tall and some short, we also vary in the cognitive abilities with which we are innately endowed. This variation—in addition to our surroundings—can influence the academic path we are likely to take. It's important to point out, however, that although Sara never performed at the top of her Piedmont class academically or went to the best universities, this didn't adversely affect her ultimate career success. Today, Sara is in her mid-thirties and a successful businesswoman, a cofounder and CEO of a prominent technology-centered advertising agency based in the San Francisco Bay area. Her agency is often lauded for its creative campaigns and "out-of-the-box" advertising style and has more work than it is able to handle. By almost any metric, Sara is indeed a success.

THINKING "OUTSIDE THE BOX"

Sara liked most subjects in school, but she was never that fond of math. Memorizing answers to multiplication and division problems or working tirelessly through algebra equations was just not Sara's idea of fun. Fortunately, in her new math class at Piedmont Middle School, she had been happy to find that a good chunk of class time was spent on logic puzzles rather than writing formulas or reciting times tables. Her teacher was on to something. He knew that having good reasoning skills was an important part of developing math competency and so he tried to hone his students' math knowledge and logic. Sara's teacher may not have known, however, that the logic puzzles he had his students complete are also used by psychologists to separate people who are low on the cognitive horsepower continuum from those near the top.

Take the following problem:

(1) **Premises:** All mammals can walk. Dogs are mammals.

Conclusion: Dogs can walk

Does the conclusion follow logically from the premises?

What about this next problem?

(2) **Premises:** All mammals can walk. Dolphins are mammals.

Conclusion: Dolphins can walk.

Does *this* conclusion logically follow from the premises?

Dolphins can't walk. But, if the two premises hold, the answer should be "yes" to both.

Almost everyone who does these problems gets the first one correct, because the conclusion that follows from the premises in problem 1 is both logical (it follows from the two premises) and believable (we do know that, in fact, dogs can walk). However, some fare better than others on the second problem. Why? The second problem not only requires reliance on logical reasoning processes, but it also requires inhibiting information about the believability of the conclusion—information that can creep into the decision-making process. One cognitive ability known to predict performance on this type of logic task is working-memory.

But is more working-memory or cognitive horsepower always better? On the one hand, research has shown that the higher your working-memory, the better you perform on academic tasks ranging from reading comprehension to math problem solving. On the other hand, some of what makes higher working-memory individuals excel on problems like number 2 above may be just what hurts them whenever thinking creatively or "outside the box" is necessary. People's ability to think about information in new and unusual ways can actually be hampered when they wield too much brainpower. This seems to be even more true the more you know about a given subject. When people with lots of baseball knowledge, for example, are asked to come up with a word that forms a compound word with *plate, broken* and *shot,* they are pretty bad at this task. Baseball fanatics want to say the word is *home* (home-plate, broken-home, home-shot?!). This isn't correct. The real answer is *glass* (glass-plate, broken-glass, shot-glass). What's interesting is that baseball fans who also have a lot of cognitive horsepower relative to their peers—those higher-working-memory baseball fans—are the ones most likely to dwell on the wrong baseball-related answer. It's as if these guys (and girls) are too good at focusing their attention on the wrong baseball information. As a result, they have trouble breaking free of their knowledge and coming up with the correct answer that has nothing to do with baseball. Baseball fanatics high in working-memory have problems thinking outside their baseball box.[1]

There are many examples of people getting stuck because they have too much knowledge and brainpower at their disposal. Take the candle box problem developed by German psychologist Karl Duncker in 1945. Duncker asked people to figure out how to attach a candle to a

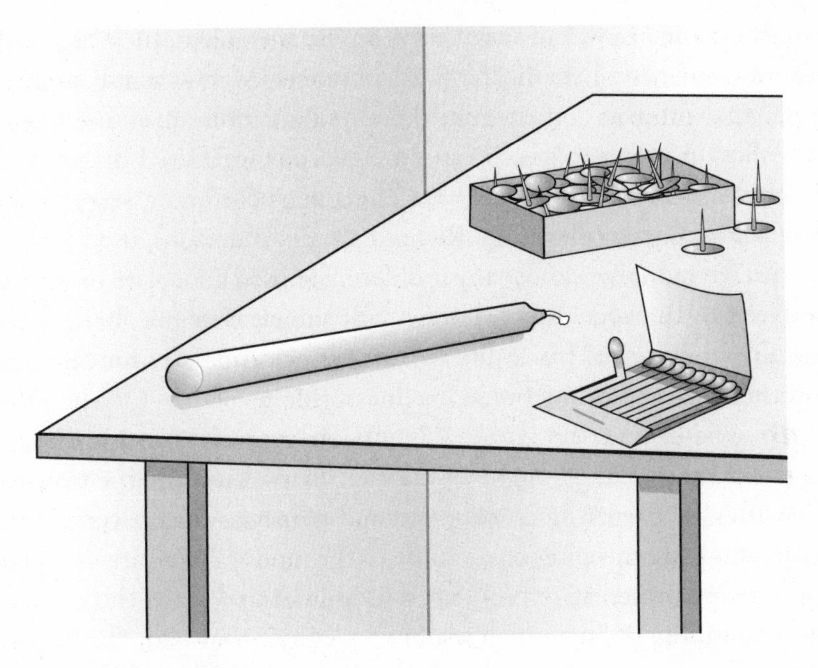

Duncker candle problem[2]

vertical wall using only a box of tacks, a candle, and a book of matches. How do you do it?

To succeed at this task, you have to realize that the tack box cannot only be used as a container, but also—if emptied—as a support. Adults have trouble seeing the tack box as anything other than a container and are notoriously bad at solving the problem. Interestingly, five-year-olds don't do as poorly. The reason is that working-memory and the prefrontal cortex in which it is housed develops with age. Adults are at a disadvantage on the candle box problem because they are too good at wielding their cognitive horsepower to focus on the normal use of a tack box—as a container for tacks. Five-year-olds, on the other hand, because they are not yet as constrained by a powerful prefrontal cortex and don't know as much about boxes and tacks, are able to come up with new and unusual ways to use the container and, as a result, they find creative solutions to the task.[3]

Of course, not all adults fall prey to functional fixedness—the inability to see new and usual ways to use an object, such as a tack box as a support. Ever heard that someone "MacGyvered" a problem? This verb was coined from the popular action-adventure television show *MacGyver,* which ran on ABC in the United States from the mid-1980s to the early 1990s. The star of the show, secret agent Angus MacGyver (played by Richard Dean Anderson), used science and his wits to solve almost any problem. He used chocolate to stop an acid leak and a paper clip to short-circuit a nuclear missile. In doing so MacGyver displayed his ability to overcome normal uses for common objects. He avoided functional fixedness.

Or what about the *Apollo 13* mission-control scientist who, in April of 1970, had to engineer a quick fix to keep their astronauts alive after an oxygen tank exploded and damaged the spacecraft the astronauts were traveling on en route to the moon? An entire life spent coming up with creative problem solutions likely helped these scientists, regardless of how much working-memory they had, think up an unusual way to keep the astronauts out of danger. To conserve power in the damaged command module, the astronauts had moved to the lunar module the command module had docked with. Unfortunately, however, there were not enough lithium hydroxide canisters in the lunar module to cleanse the carbon dioxide that the astronauts themselves expelled (scientists hadn't planned that the entire mission team would need to spend so much time in the lunar module). There were the canisters used in the command module, but they were incompatible with the system in the lunar module. The scientists on the ground came up with a solution to connect the command-module canisters to the lunar module's system using plastic bags, cardboard, and duct tape. In short, the scientists were able to come up with a new and unusual way to use common and available objects to help their astronauts breathe.

Two strings hang from the ceiling but are too far apart to allow a person to hold one and walk to the other. On a table under the strings are a book of matches, a screwdriver, and a few pieces of cotton. How can you tie the strings together?

I asked two siblings I know, a second grade boy named Dean and a seventh-grade girl named Isabella, how they would solve this problem and, after careful thinking, they each came up with a different answer. Isabella said that one possibility was to use the screwdriver as a weight that could be tied to one of the pieces of string. She then proposed that you could swing the screwdriver and string in a pendulum motion so that you could hold on to the other piece of string and catch the pendulum when it came within reach. Dean thought this was a good idea, but he had simply decided to stand on the table so that one could grab both strings and tie them together, a very simple solution.

Whenever people can't see the screwdriver as a pendulum, but only as a tool for screwing things in, they are stuck in a mind-set that doesn't allow them to find creative or outside-the-box problem solutions. Our seventh grader, Isabella, was able to avoid this type of fixation. Her advanced prefrontal cortex (at least compared to that of her second-grade brother), however, didn't allow her to see an even simpler solution to the problem, which involved just standing on the table.

When people highest in working-memory solve a problem like the string task, which requires thinking about the situation in an unusual way, they often struggle to find a quick and easy solution. Many adults never think of the screwdriver as a pendulum and even fewer think to stand on the table. High-powered folks often opt for the most difficult way to get through tasks and, even when they do come up with the correct answer in the end, they waste a lot of time and energy doing so.

A few years ago, my Ph.D. student Marci and I showed this high-power disadvantage very clearly. We asked college students to solve a series of math problems known as the Luchins water jug task and then looked at how the students went about finding solutions to the problems as a function of whether they were lower or higher in working-memory.[4] The Luchins task presents a picture of three water jugs of various sizes and a goal quantity of water that the problem solvers are supposed to end up with.

People are asked to come up with a mathematical formula that uses the size of each jug (the number below the jug) to obtain the goal water quantity. Folks have as much water as they need to do this task,

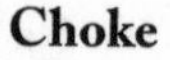

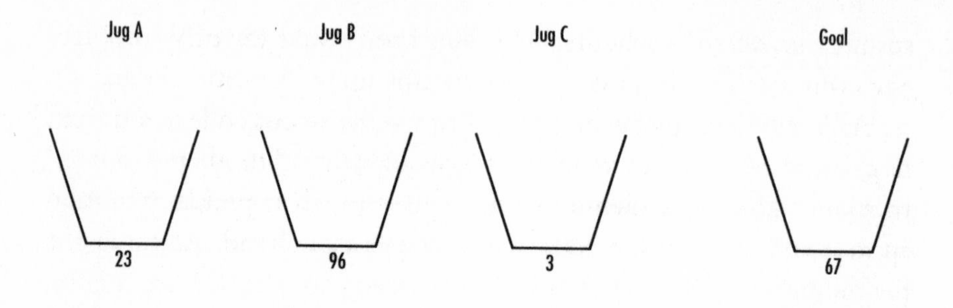

but are asked to work under one important constraint: they must use the *simplest* strategy possible to come up with the answer.

Marci and I gave roughly a hundred college students the same six water jug problems to solve. The first few problems were only solvable using a pretty difficult multistep strategy. For instance, to solve the above example, first, you have to fill up Jug B with water (96 units), then you pour it into Jug A (23 units), and then you pour what is left of Jug B (that is, 96 − 23 or 73 units) into Jug C . . . twice (73 − 3 − 3 = 67 units, the goal water quantity). The official formula to solve the above problem and the first several problems we gave students is "B − A − 2C" (that is, 96 − 23 − 3 − 3). This formula works for the next problem as well (49 − 23 − 3 − 3 = 20 units, the goal water quantity). But there is another way to solve the problem below that is much simpler.

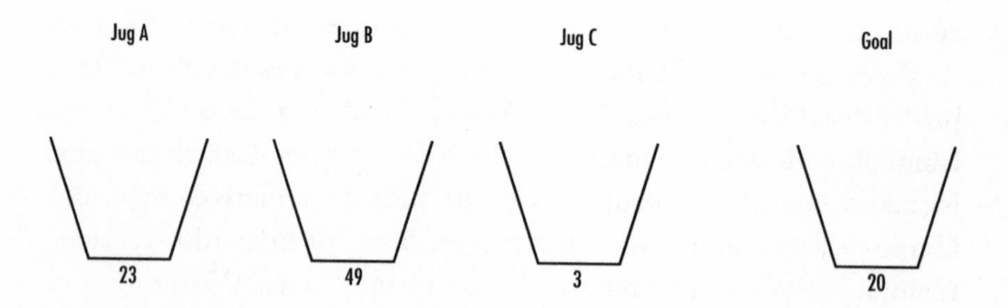

It is "A − C" (or 23 − 3). Marci and I were interested in finding out whether people were able to find this shortcut when it occurred (remember, they were told to solve the problems using the simplest strategy possible) or whether they continued to use the difficult strategy even though a less intense option existed.

As we suspected, the more working-memory our college students had, the less likely they were to find the simpler solution. The higher-working-memory students missed the economical problem-solving approach. The low-powered students, on the other hand, went straight for the easy solution.

Why would more working-memory translate into a higher likelihood of missing the easy solution on Luchins water jug task? Or, for that matter, why would it translate into more difficulty in breaking away from a baseball-related solution in the word association task or figuring out that a tack box, if emptied, can be used as a support device?

Focusing your attention on the most important information and ignoring less relevant input is something that people higher in working-memory are very good at. In many situations, this ability to control your focus of attention can be advantageous. This is certainly true in the second logic puzzle presented above—when you have to ignore the believability of the statement that "Dolphins can walk" to correctly answer that the conclusion does follow logically from the premises. But this isn't always the case. A narrow focus of attention can prevent people from detecting alternate solutions to a problem. This narrow focus can even hamper your ability to notice unexpected events around you.

Consider one of the most memorable occurrences in college football history. The date was November 20, 1982, and the location was Memorial Football Stadium at the University of California. This is where one of the greatest rivalries in college football—the Big Game—was being played out between the University of California Golden Bears and the Stanford University Cardinals. With four seconds left on the clock, Stanford kicked a field goal to take a 20–19 lead. Most everyone thought the game was over, including the Stanford band, who had decided to take the field in preparation for their team's ensuing victory, even though the last play was still unfolding.

In one of the most unbelievable football plays of all time, the Golden Bears returned the Stanford kickoff using five lateral passes to come away with a 25–20 victory.

The following year *Sports Illustrated* ran a twelve-page spread on the play, calling it "The Anatomy of a Miracle." The play, *Sports Illustrated*'s experts concluded, was completely legal, despite the fact that the Stanford band had taken the field before the game was complete. Band members hadn't even seen the California team coming, as evidenced by the fact that Cal's Kevin Moen, who had charged the end zone after having caught the ball at the twenty-five-yard line, knocked over Stanford trombone player Gary Tyrrell, who was standing on the field with no idea as to what was going on.

The trombone player, Tyrrell, might have missed the football player barreling down on him because he was too reliant on working-memory and his prefrontal cortex. When the prefrontal cortex takes over, other brain regions—such as sensory and motor areas—have less room to chime in. These sensory and motor brain areas are quite sensitive to unexpected occurrences in the environment, whereas the prefrontal cortex, on balance, works to uphold people's expectations about a situation. Those who rely too heavily on their prefrontal cortex may miss unexpected events precisely because they are not getting as much benefit from the brain areas best equipped to process the outside world.

Think back to the last time you were at a party. That's right, a party. Your ability to overhear your name being spoken across the room—not when someone is shouting it to get your attention but instead when someone is talking about you behind your back—increases as your cognitive horsepower decreases. This ability to detect your name in a conversation you are not really listening to is called the *cocktail party* effect.[5] Yes, it's actually a psychological term. People lower in working-memory show a stronger cocktail party effect than people with higher working-memory because low-powered folks have a hard time focusing in on just one thing and instead are always paying attention to a little bit of everything. Hence they hear their name even when they are not listening for it. This ability to pick up on unexpected events might have been just what Stanford trombone player Gary Tyrrell was missing.

So, acquiring new information—like knowing when people are chatting about you across the room at a party—is sometimes best accomplished with less cognitive horsepower. Admittedly, eavesdropping ability is generally not thought of as a great academic asset. But not being able to focus your attention completely *is* handy for learning some skills that are important in education and job settings—such as language. Moreover, it's a common assumption that people with less working-memory (for instance, people with Attention Deficit Hyperactivity Disorder, ADHD, in which working-memory deficits play a major role) are always at a disadvantage in important performance settings. However, there are certain activities where less is more, as you will see below.

LEARN LIKE A KID

People who speak two languages from a very young age have a level of second-language proficiency that, on average, far surpasses that of people who begin to study a language later in life—no matter how hard they hit the language books as adults. One of the reasons kids may be so good at picking up languages is that cognitive horsepower develops with age. Because children have lower working-memory capabilities than adults, this actually aids their acquisition of foreign tongues.

To learn a language, you must be able to correctly select many different pieces of information from a stream of conversation. This includes the words that are spoken and their particular combinations. It also includes the subtle alterations in words that change their meaning—such as adding an *s* to the end of a word to change it from singular to plural form. By analyzing the errors that people make when learning a second language, psychologists have discovered that adults make some errors that kids don't precisely because adults have too much working-memory at their disposal.

For instance, adults are more likely to treat whole words as units—bringing letter combinations, words, or phrases that often appear together into a new context, even when these combinations aren't appropriate. So, for example, adults are likely to keep the *s* on a word that is supposed to denote a singular because they have heard it previously in plural form. Children, on the other hand, are better able to pick up the individual pieces of the language they are exposed to, which helps them use language pieces flexibly and correctly. Because kids can only attend to bits and pieces of what they hear, this helps them pick up the particulars of a language.

Researchers Alan Kersten and Julie Earles have shown that adults learning a made-up language learn word meaning and word rule use better when they are initially presented with only individual words and then later with more complex sentences.[6] Adults do less well when they are introduced to all the complexities of the language from the outset. Small segments allow the adults to process the language as if their working-memories were more limited in the first place, as if they were under the same prefrontal developmental constraints as children, and this in turn helps them learn language more quickly.

Of course, cognitive horsepower is beneficial rather than detrimental to performance in a lot of situations. When you are trying to solve a complicated math problem in your head, such as figuring out which item at the grocery store is a better buy when one is priced per pound while the other is priced in gallons, the more working-memory you have the better. But when you need to think creatively or outside the box, or engage in flexible problem solving, the higher your cognitive horsepower (or if you are unable to dampen your working-memory on command) the more difficulty you may have. People with lower working-memory will generally do better.

Just in case I have not yet convinced you that less can sometimes be more, let me tell you about one additional study in which people with damage to the prefrontal cortex—the brain area that houses working-memory—outperform those without brain damage. A group of Italian scientists asked patients with lateral prefrontal cortex damage that had occurred through injury or stroke to solve some unusual math problems. They also asked a group of healthy adults without brain damage to solve the same problems.[7]

As a warm-up, both the patients and the healthy adults were given arithmetic problems like the one below. The problems were constructed completely out of matchsticks and the goal was to make the false statement true by moving a single stick:

IV = III + III

To come up with the answer to the problem above, you have to move the leftmost stick of number IV to the immediate right of the V to make:

VI = III + III

More than 90 percent of the healthy adults and roughly the same proportion of patients with prefrontal cortex damage got problems like the one above right. This is not so surprising, because the matchstick to be moved is pretty obvious.

However, when people were asked to solve the problem below where they had to think a bit differently about what each matchstick meant, only 43% of the healthy adults solved it. This is in contrast to 82% of patients with prefrontal cortex damage who got the problem correct.

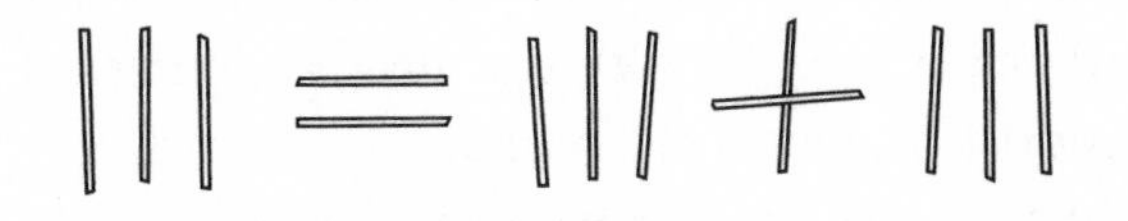

Give up? The answer to the above problem involves changing the plus sign by rotating its vertical matchstick ninety degrees into an equal sign. Crucially, this action transforms the starting equation into a tautology:

Patients with prefrontal cortex damage were able to look at the problem in an unusual way and their new perspective enabled the patients to see the potential to turn the plus sign into an equal sign. Adults with their working-memory intact, on the other hand, were too good at focusing in on the normal constraints of the math problem and couldn't see the unusual solution that involved changing the operator.

MORE OR LESS?

What should you make of all of this? On the one hand, having more working-memory is often beneficial. Indeed, folks who score highest on working-memory tests also score at the top on most measures of academic achievement. On the other hand, we just saw several examples where the ability to focus in on one piece of information and ignore others—an ability that is at the heart of working-memory—is actually detrimental to creative thinking, language learning, and "outside-the-box" reasoning.

Just think about Sara, our seventh grader whose parents uprooted her from Oakland to Piedmont to expose her to the best-of-the-best public schools. Sara never shined on standard metrics of academic achievement such as school grades and test scores and would have likely been positioned on the lower end of the working-memory continuum. Even so, she was creative and always found new and unique ways to look at the problems she encountered, which led to her eventual success in advertising—a profession where thinking "outside the box" is not only important but necessary. So, is it better to be higher in cognitive horsepower or not? The answer is, yes . . . and no. The key is to have brainpower at your disposal, but to be able to "turn it off" in situations where it may prove disadvantageous.

Psychologists at Louisiana State University seem to have found some clues for doing this, at least for language learning. Adults are better at acquiring a new language—that is, adults look more like kids with underdeveloped prefrontal cortexes—when they are distracted and not concentrating too hard on what they are learning.[8] The researchers taught college students a modified form of American Sign Language (ASL) in which the students learned to sign simple sentences such as "I help you" and "You help me." To do this, students watched a videotape in which the sentences were shown, first in English subtitles, and then by a person signing them. Some students learned the sentences with nothing else going on. Other students watched the ASL learning videos and were asked to do a distracting task, involving counting the number of high-pitched tones played over the ASL video, at the same time. At the end, all the students tried to produce new sentences made up of some of the signs they had learned.

Adults are better at acquiring a new language—that is, adults look more like kids with underdeveloped prefrontal cortexes—when they are distracted and not concentrating too hard on what they are learning.

As it happens, the students who learned the ASL sentences with no diversions were worse at signing the new sentences than the students who had been distracted by the tones during learning. The nondistracted students had problems producing the individual signs from the sentences they had learned in new ways. Keep in mind that producing new word or sign combinations is what language is really all about. Being distracted during language learning, on the other hand, forced students to learn the individual signs because they couldn't hold the entire sentences presented to them in working-memory. Having less working-memory at their disposal helped students generalize what they had learned to new combinations. These distracted students were functioning more like children and, as a result, they were better language learners.

In a similar vein, my research team and I have found that highly skilled golfers are more likely to hole a simple three-foot putt when we give them tools to turn off their prefrontal cortex.[9] Highly practiced putts run better when you don't try to control every aspect of performance. Limiting the working-memory and conscious control you have to devote to skill execution can increase your possibility of success. Having a golfer count backward by threes, for example, or even having a golfer sing a song to himself uses up working-memory that might otherwise fuel overthinking and a flubbed performance.

Having a golfer count backwards by threes, or even having a golfer sing a song to himself uses up working-memory that might otherwise fuel overthinking and a flubbed performance.

The idea that more cognitive horsepower or conscious attention is *always* better doesn't hold—whether you are learning a new language or taking a putt that you have made hundreds of times in practice. Indeed, in some cases it's better to put your working-memory to sleep all together, literally. Rapid eye movement sleep (or REM sleep) is characterized by decreased activation of the prefrontal cortex and increased activation of brain areas such as sensory cortex. Recent studies have shown that, after REM sleep, people are better able to see connections between seemingly disparate pieces of information.[10] One reason this may be the case is that working-memory and the prefrontal cortex stop working so hard, allowing for the formation of what were, at first glance, nonobvious connections.

Of course, being able to flex your working-memory on demand is easier said than done. One of the main reasons that people choke under pressure is that they are not using their working-memory in the right way: they are either paying too much attention to what they are doing or not devoting enough brainpower to the task at hand. In the chapters to come we will explore exactly how pressure derails performance and what you can do to ensure that you are using working-memory optimally—especially when your performance counts the most.

Before we turn to failures of working-memory under pressure, however, let's address how individual differences in working-memory come about in the first place and what you can do to improve your own so that you have it when you want to use it.

GENES AND HORSEPOWER

Starting out at a new school—especially in seventh grade—is no picnic. Fortunately for Sara, she made a tight-knit group of friends who eased her transition into Piedmont Middle School. Five of them, including a pair of identical twins, a pair of fraternal twins, and Sara herself, were pretty much inseparable that year.

Identical twins share all of their genes; fraternal twins share only about half of their genes. This was pretty obvious in the twins' physical appearances. The identical twins were hard to tell apart if you didn't know them well. In contrast, one fraternal twin was taller than her sister and each had a different shade of blond hair.

When she was in junior high, Sara would not have thought of running a psychology experiment. She was way more concerned about boys, clothes, and sports. Yet Sara and her group of friends had the makings of a very interesting study concerning how people develop different amounts of working-memory or cognitive horsepower. Sara's identical-twin friends grew up in the same environment—they shared the same house, the same parents, and each was given the same educational opportunities. The fraternal twins were reared in a common situation as well. This means that if genes do play an important role in creating differences across people, the identical twins' cognitive abilities should be more similar to each other than the fraternal twins' ability would be. After all, identical twins share twice as many genes as the fraternal twin pair.

With only one observation of each type of twin to go on, Sara would have been hard-pressed to come to any definitive conclusions about individual differences and their origins. But a group of researchers at the University of Colorado have turned Sara's potential experiment into reality. Over the past several years, psychologist Naomi Friedman and her colleagues have been working with teenage twins to find out whether the cognitive abilities of identical twins look more similar to each other than the abilities of fraternal twins do. Their goal is to figure out how genes shape differences across people in intellectual and academic ability. To do this, the Colorado researchers selected hundreds of pairs of identical and fraternal twins who were raised together in the same environment, then asked them to perform tasks to get at their working-memory.

As a reminder, working-memory is more than just storage; it also reflects your ability to hold information in memory while doing something else at the same time. Controlling your focus of attention is key here so you don't forget or get confused about what you are trying to remember in the first place. This is why working-memory and attention are often talked about in the same breath—working-memory involves being able to attend to some things and ignore others so that you can keep the information you want to remember in mind. Differences in working-memory among individuals account for between 50 to 70 percent of variation in general intellectual ability. In short, working-memory is one of the major building blocks of IQ. So researchers are interested in how we acquire it in the first place.

> Working-memory is more than just storage; it also reflects your ability to hold information in memory while doing something else at the same time.

The concept of working-memory is admittedly a bit fuzzy. However, it becomes much clearer when we take a look at some of the tasks used to gauge it. As you may recall from chapter 1, a common task used to assess working-memory is the reading span or RSPAN, task. In this task, individuals are asked to remember a list of words while deciding whether a series of unrelated sentences makes any sense. The essence of this task—holding information in your head while having to combat other stuff from weaseling its way into memory—is quite similar to some of the tasks the Colorado research team had their twins do.[11]

In one task, the *letter-updating* task, twins were presented with a string of letters on the computer, one at a time, and for only a few seconds each. As this was going on, the twins were asked to recall the *last* three letters they had just seen. So, if you see the letters "*T . . . H . . . G . . . B . . . S . . . K . . . R,*" then you are supposed to say aloud as each letter appears, "T . . . T-H . . . T-H-G . . . H-G-B . . . G-B-S . . . B-S-K . . . S-K-R." This task requires that you continuously pay attention to the new letter that is appearing and that you appropriately update the letters you are holding in memory by replacing the oldest, no longer relevant letter with the newest letter coming into the list. You have to draw on your working-memory so that you do not become completely confused by this task.

The Colorado research team found that the identical twins looked more similar to each other than the fraternal twins did on the letter-updating task and on most of the measures of cognitive ability that the researchers used. Their results are not too far off from what Sara remembered about her seventh-grade friends. Her identical twin friends always led the class, and competed for top grades, test scores, and teacher evaluations. In fact, their marks in school from kindergarten to seventh grade were almost . . . identical. The fraternal twins were somewhat more variable. Thus, from the University of Colorado researchers' findings and Sara's casual observations of her friends, it seems that the more genetically similar two people are, the more similar their cognitive capacities. In other words, some of the differences between people in cognitive ability seem to be genetic in origin.

You might be wondering how this innate influence actually comes about in the first place, and you are not alone. In the past several decades, scientists have become increasingly interested in the relations between genes and cognitive functioning. Although there is still a lot of work to be done on this front, scientists are making progress. For instance, researchers have found that the neurotransmitter dopamine (a chemical that relays signals from one brain cell to another) is involved in enhancing your ability to update information in memory and also contributes to your ability to focus squarely on the task at hand without being distracted. These memory and attention processes may sound familiar, because they are very similar to what you need in order to ace the letter-updating and RSPAN tasks described thus far. At any rate, molecular genetic studies have found that different people have different versions of the gene (called COMT) involved in the metabolism of dopamine.[12] Because the different versions of the COMT gene (the two main forms are termed Val and Met) break down dopamine more or less efficiently, they are thought to be linked to individual variation in cognitive horsepower.

There is some exciting new work that shows that cognitive horsepower or working-memory—once thought to be an immutable trait that was completely heritable in origin—can be altered with the right type of brain practice.

Even though the evidence regarding the genetic basis of cognitive functioning is compelling, it's important to remember that the environment plays a large role in shaping success. Indeed, there is some exciting new work that shows that cognitive horsepower or working-memory—once thought to be an immutable trait that was completely heritable in origin—can be altered with the right type of brain practice.

TRAINING COGNITIVE HORSEPOWER

Jason always led the way on the football field, but often struggled in school. In elementary school, several teachers noted that Jason had a lot of trouble sitting still in class. Jason had always been a somewhat hyperactive kid, which his parents had accepted as part of his rambunctious personality. But when it became apparent that Jason was unable to pay attention to his teachers for prolonged periods of time or to keep his attention squarely focused on the tests he took,

Jason's parents took him to see a psychiatrist, who diagnosed him with ADHD (attention deficit hyperactivity disorder).

At the time, Jason's parents were frustrated because they felt they didn't have many ways to deal with the diagnosis. Not wanting to turn immediately to the medications that were available, they instead opted to buy Jason a new pair of football cleats in the hopes that he might work off some of his extra energy on the playing field. This tactic is not unusual. Take fourteen-time gold medal Olympian swimmer Michael Phelps. Like Jason, Michael was diagnosed with ADHD in elementary school, which is one of the reasons his mother put him in the pool—so that Michael could channel his energy into something productive. Now psychologists are discovering that, just as practice can improve your swimming stroke or soccer kick, brain practice can be used to lessen the symptoms of ADHD.

The American Psychiatric Association characterizes ADHD by inattention, impulsive behavior, and hyperactivity. Central among the deficits that accompany ADHD is an impairment in working-memory. For a long time, working-memory was considered to be immutable—meaning that it couldn't be trained or practiced. Some individuals had more working-memory at their disposal, and some folks, such as children with ADHD, had less. Fortunately, however, in recent years, researchers like Torkel Klingberg at the Karolinska Institute in Stockholm, Sweden, have made it their mission to abolish the notion of a fixed working-memory capacity. Klingberg has been especially interested in whether children with ADHD can improve their focus of attention and lessen their hyperactive tendencies with brain training. His logic: if impaired working-memory is a core deficit in ADHD, then training working-memory should decrease ADHD symptoms.

Klingberg and his colleagues have conducted a number of studies to test his idea, but let me describe just one to you that really captures the malleability of working-memory quite well. In this study,[13] Klingberg and his team randomly assigned children diagnosed with ADHD to one of two groups: a treatment group or a placebo group.

In the treatment group, the kids went through an intensive working-memory training program that involved a variety of verbal and spatial tasks designed to exercise kids' memories for five weeks in a row. In one task, the treatment group kids saw letters—one by one—on a computer screen. They were to remember the letters as they appeared and then to recall them back to the experimenter in the exact opposite order in which they had been originally presented. This type of backward memory task is quite hard because you have to keep track of what is presented to you and reverse it in your head. This reversing is the working part of working-memory. Critically, as the children in the treatment group got better and better at the backward memory task, the difficulty—that is, how many items they had to remember and reverse in mind—increased. In essence, the training was always pushing the kids to work their working-memory a bit more.

Children in the placebo group also performed a variety of activities on the computer that were similar to those done by children in the treatment group. However, when kids in the placebo group performed the backward memory task described above, they only had to remember a few items and the number of items never increased. Children in the placebo group had much less of a working-memory workout.

Not surprisingly, kids in Klingberg's treatment group got better on the working-memory tasks they trained on. But these kids also improved on other attention and reasoning tasks that they had *not* practiced. Even more impressive is that, in a similar study after the training, kids in the treatment group were able to sit still longer and were less hyperactive than their placebo-group counterparts. Working-memory training led to a decrease in the symptoms associated with ADHD.

Importantly, Klingberg designed his experiment as a double-blind study. The children, parents, and even the psychologists administering the tests before and after training did not know which version of the computer program (that is, the treatment version or the placebo version) on which the kids had practiced. This ensured that neither the parents' nor the experimenters' input nor even the expectations of the kids involved in the study could explain these results. It's no secret that Michael Phelps benefited in the sporting world from his intensive training in the pool. Klingberg's work suggests that Phelps might have benefited in the classroom from a brain-training regimen as well.

It's not just kids with attention deficits who can improve their core cognitive abilities with training. Michael Posner, a neuroscientist at the Sackler Institute in New York who has spent most of his academic career studying the concept of attention, recently found some interesting ways to train brainpower in normal children in order to enhance their learning and performance in school. Much like Klingberg, Posner treats the brain like a muscle that needs training in order to grow.

Posner and his colleagues asked kindergarten kids to do things like learn how to use a joystick to control the movement of an animated object on the computer screen and to predict where that object might go given its initial trajectory. Many of these tasks were patterned after those once used to gear up rhesus monkeys in the United States and Russia for space travel. Just as in Klingberg's work, each exercise the kids did progressed from easy to more difficult so that the children were pushed to expand their attention and memory skills with practice.[14]

Sure enough, Posner's brain training led to improvements on the tasks that the kids practiced. Even more striking, however, was what Posner and his research team found when they looked at the kids' brain functions after practice. Children's neural activity in frontal brain areas like the anterior cingulate cortex, which, among its many functions, is involved in controlling and shifting attention, resembled the activity of adult brains when performing difficult tasks. Because the functions of these brain areas develop with age, Posner's work suggests that attention training may be one way to jump-start the developmental process.

This type of workout for the mind seems to help adults as well. Recently, Klingberg asked normal twenty-somethings to spend several weeks training on working-memory tasks similar to the ones that his ADHD children had pursued—for example, remembering letters and then repeating them back in the reverse order of how they were presented.[15] After training, adults were not only better at the tasks on which they had trained, their improvement generalized to several attention and reasoning tasks that they had not practiced. Klingberg used fMRI to see how this working-memory improvement came about, and found that brain areas that support cognitive horsepower, such as the prefrontal cortex, showed more activity after training than before training. Just as after a period of weight training people are able to lift heavier weights than they did before they started to build muscle, these changes in prefrontal activity seem to indicate that the brain can do more work with practice.

The prefrontal cortex had undergone considerable expansion throughout evolution. Our ability to control attention and hold information in memory with our powerful prefrontal cortexes may be one of the things that sets us apart from other animals. We generally think of people who are naturally better able to harness their prefrontal cortex as being higher in cognitive horsepower. Now, however, we know that this prefrontal cortex variability is not fixed and that cognitive ability improves with training.

HURRAY FOR VIDEO GAMES!

Admittedly, not everyone has access to the types of sophisticated training regimens that Michael Posner and Torkel Klingberg use to help folks strengthen their cognitive horsepower. The good news is that you can flex your working-memory in several different ways. Playing action video games, for example, can improve your brainpower. That's right, spending several hours a week playing games like Grand Theft Auto, Half-Life, or Halo improves core cognitive abilities that extend well beyond the computer screen.

In one study, college students with little previous video game experience were asked to play the popular video game Medal of Honor for ten days in a row.[16] The game takes place during the end of World War II and players take on the role of Lieutenant Jimmy Patterson, who is recruited by the OSS (Office of Strategic Services) to help the

United States advance their mission. Players must accomplish objectives such as to destroy enemy positions and kill as many German soldiers as possible. These wartime objectives require working-memory. Players must constantly move their attention from one aspect of the game to another so that they do not miss incoming enemies or new developments. While doing this, they must also keep their mission goals updated and fresh in their minds. In short, players must juggle a number of tasks at once and, to succeed, they can't drop the ball on any front.

After playing Medal of Honor an hour a day for ten days, college students showed improved memory and attention abilities on a number of different tasks. Importantly, people improved even on tasks that they had not directly practiced. The better that people got at Medal of Honor, the more their attention and memory skills outside the game skyrocketed.

So, parents, before you take your kid's Nintendo DS away for good, you might want to think about the potential benefits of some video game play. Keep in mind, however, that these benefits occurred after only an hour of play a day. Eight hours a day, every day, will likely have diminishing returns in upping cognitive horsepower.

Nonetheless, some video game play—especially if the game helps you practice important cognitive skills—can be good. The Israeli Air Force thinks so and, in the mid-1990s, psychologist Daniel Gopher asked Israeli Air Force cadets to play a video game that he had helped to develop.[17] Called Space Fortress, it exercises memory and attention skills by requiring players to maneuver an aircraft, using a joystick, through a frictionless, hostile environment, firing missiles to defeat the enemy while at the same time avoiding being shot oneself. After a mere ten hours of game play, an hour a day spread over ten days, the cadets showed almost a 30 percent improvement in their actual flight performance. The Israeli Air Force was so amazed with these results that Space Fortress is now a permanent part of their flight-school training program.

In recent years, the use of Space Fortress has extended beyond the battlefield to basketball, where players also have to make numerous important strategic decisions. Several college and professional coaches are now using a version of the game as part of their team training. Many of the 2006 NCAA champion Florida Gators used it and *Slam* magazine, one of the premier magazines for NBA and college basketball enthusiasts, calls it a "workout for the mind" that is not to be missed. Of course, in training cognitive horsepower, Space Fortress may be improving ballers' skills and performance on the court and in the classroom.

A LOOK AHEAD

Despite the fact that people vary in their innate abilities, training plays a large role in where our abilities lead us. Yet even if you don't measure at the high end of cognitive horsepower, this doesn't mean you can't excel in academics or in the business world. Indeed, University of Michigan psychologist Priti Shah and her colleagues have shown that college students diagnosed with ADHD are actually better than non-ADHD students at coming up with creative solutions to problems. ADHD students, for example, are able to generate more unusual uses for common objects than students without ADHD. One way this might play out in the business world is that ADHDs, and people lower in working-memory in general, will be successful at coming up with creative applications (think iPhone apps) for existing technology. Shah thinks this creativity stems, in part, from ADHD students' lack of ability to inhibit information from creeping into the mind, which leads to more divergent thinking.[18] Thus, even though you can train your working-memory, including your inhibition abilities, less is sometimes more.

Nonetheless, despite recent work showing that cognitive horsepower is malleable and that what matters most is your ability to use your working-memory when you need it and dampen its influence when you don't, some scientists still try to measure average or mean differences in IQ, for example, between groups of people—demarcated by sex or ethnicity, for instance. These perceived (and misperceived) differences between the sexes or races in performance are then used to justify sexist or racist policies in education and employment, which of course creates heated debates.

In recent years, naturally occurring ability differences have received renewed attention—sparked in part by some provocative comments made by then Harvard president Larry Summers in 2005. In the next several chapters we look at what research reveals about genetic and environmental factors and about how opportunities for training and practice may contribute to the sex and ethnic divide—especially when this divide is measured by high-stakes tests. Interestingly, the very act of asserting group differences in cognitive functions such as working-memory based on sex or race can create a stressful situation where the individuals being pigeonholed are likely to perform below their abilities.

CHAPTER FOUR

BRAIN DIFFERENCES BETWEEN THE SEXES

A SELF-FULFILLING PROPHECY?

On a cold winter afternoon in the middle of January 2005, Lawrence (Larry) Summers took the floor to give a lunchtime talk at the National Bureau of Economic Research (NBER) conference on diversifying the science and engineering workforce. Held in Cambridge, Massachusetts—the home of Harvard University—the conference was convenient for Summers to attend because he was at the time president of Harvard.

Although the NBER conference was centered on an important topic, diversity in science and engineering, the work being presented to an audience mostly of university administrators and faculty was not intended to make news headlines. Summers had warned at the outset of his talk, however, that his comments were going to be provocative. In fact, he stated up front that he was *not* at NBER to speak on behalf of Harvard University, but to speak "unofficially" about diversity. He asked why women make up a minority of the tenured positions in science and engineering at top universities and touched on several reasons for their underrepresentation in STEM (science, technology, engineering, and math) fields. On answering this question, he interpreted data in ways that his audience immediately reacted against as sexist.

Summers didn't dwell on whether or not, on average, men's academic ability is superior to that of women's, but he stated that men are more *variable* in their intellectual capabilities than women are.[1] In other words, more men achieve higher levels of aptitude at the top of these disciplines. Because more variability generally means having more people at the extreme ends of a distribution, Summers's point was, in a nutshell, that there are simply more men than women at the top end of the math and science talent pool, so they get more of the jobs. To support his claims, Summers explicitly mentioned research that had documented the sex ratio in the top 5 percent of twelfth graders, where he pointed out that boys outnumber girls by at least 2 to 1. One might criticize Summers for cherry-picking data that supports his arguments, but research also shows that, at the highest levels of performance on the SAT-M, boys also outnumber girls.[2] This is true on the Advanced Placement (AP) tests that students take in high school to earn course credit for college, as well. Although girls complete more AP tests overall than boys, more boys take the calculus test and a higher percentage of boys than girls have clear passing scores.[3] Summers's explanation for why women have not advanced to the top in math and science is based on the idea that there are simply more men with high abilities and skill available to fill the most prestigious jobs. He downplays the socialization of the sexes that could lead to this disparity.

You might be interested to know that Summers's argument isn't a new one. This greater variability hypothesis can be traced as far back as Charles Darwin, who, in *The Descent of Man,* suggested that men vary more in their physical characteristics than women do. It's a short hop from physical characteristics to intellectual ones and, by the early 1900s, many psychologists were asserting that men are more different from each other in terms of their intelligence than women are from each other. The "natural" end result of this line of thinking is that there will be more men than women at very low levels of intelligence, but importantly for Summers's assertions, there will also be more men than women at the highest levels.

Because of these perceived sex differences in variability, in 1906 American psychologist Edward Thorndike suggested that the education of women for professions where "a very few gifted individuals are what society requires, is far less needed than for such professions as nursing, teaching . . . where the average level is the essential."[4] After all, if you have a higher percentage of men at the highest ability extremes, then maybe it makes sense to focus your attention solely on men to find the most outstanding individuals.

Acting as a modern-day Thorndike, Summers argued that girls and women are more likely to display academic ability that is of an average level while boys and men are more likely to be below or above average, at the ability extremes—especially in math and the sciences. Because there are simply more men at the very top of the talent pool, Summers asserted, men are more likely to be tapped to fill big-time science jobs. According to Summers, the disparity between men and women in academic research institutions reflects naturally occurring variation.

Summers is often lauded for his speeches and his intellect, and has been called one of the greatest economic minds of his time. Yet his talk that afternoon didn't engender the positive reactions he was used to. On the contrary. Nancy Hopkins, a biologist at the Massachusetts Institute of Technology, walked out in the middle of Summers's remarks and later said that she might have "thrown up" in disgust if she had not left. Several other audience members, ranging from chancellors to deans at prominent universities across the country, went on the record saying they were deeply offended by Summers's argument.[5]

In addition, I would question whether someone can speak "unofficially" about differences between the sexes that he is calling innate while presiding over a world-class research institution. Given that Summers stepped down as Harvard's president shortly after his NBER remarks, others would join me in asserting that one cannot.

Yet you have to wonder what a Harvard president used to push his viewpoint and what research backs up the idea that there is "different availability of aptitude at the high end." Why would someone in Summers's position feel comfortable making these assertions?

The notion that there are naturally occurring differences among groups is not new. One of the most notorious peddlings of this viewpoint comes from Richard Herrnstein and Charles Murray's *The Bell Curve*,[6] in which the authors put forth the view that IQ differences between the races are substantially genetic in origin. However, as we will see below, there is a lot of evidence that the performance gap between boys and girls in math is narrowing, a trend that exists in black/white achievement differences as well. It's hard to explain these rapid changes in the achievement gap if you endorse the idea of fixed, naturally occurring differences across groups. The changes are easily explained by both girls' and minorities' increased educational opportunities over the past several decades. Of course, being knowledgeable and being able to show what you know on a high-stakes test are not the same thing, and as it happens, merely bringing up the viewpoints that Summers did (and which are echoed in terms of racial differences in *The Bell Curve*) is enough to send the most capable female and minority students spiraling downward. So scientists still need to work out when and where performance differences between the sexes and racial groups show up and, when they do, why they do. Let's focus on the issue of differences between the sexes below.

In the early 1980s, Camilla Benbow and Julian Stanley published a paper in *Science*, one of the most prestigious scientific journals in the world. Their paper details the results of a project that tracked the performance of almost forty thousand students across the country, around the age of thirteen, who took the SAT.[7] The SAT formerly stood for Scholastic Aptitude Test. Now it carries no particular name at all, but it is one of the major exams used to evaluate students for college admission in the United States and abroad. One reason for this name change is that the SAT doesn't really assess students' *aptitude* for the material they will learn in college. Rather, most of the items on the SAT are based on reasoning about topics covered in the high school curriculum. So why were Benbow and Stanley interested in the performance of thirteen-year-olds on a college admission test?

The researchers were involved in a math talent identification program designed to identify the best and brightest math minds at an early age. Their goal was to use the math portion of the SAT (the SAT-M) to find kids operating at the top in math, in order to provide them with the support and training they needed to excel in middle school math, high school math, and beyond. Of course, having access to test scores from a math talent identification program also gave Benbow and Stanley a unique chance to look at boys' math performance relative to girls'.

They found that, by age thirteen, sex differences existed in SAT-M scores. But even more striking, the researchers discovered that these sex differences were especially pronounced at the high end of the scoring distribution. In fact, for those students scoring 700 or above on the SAT-M (the 95 percentile for twelfth-grade college-bound males), boys outnumbered girls 13 to 1.

Based on math concepts typically taught in the first few years of high school, the SAT-M is designed to measure the mathematical reasoning of eleventh and twelfth graders. Most of the thirteen-year-olds taking the SAT-M as part of the talent identification program had not yet been exposed to the math on which they were being tested—either on their own or in a math class. So Benbow and Stanley thought that high scores from these youthful test takers would reflect general mathematical ability rather than what they had learned thus far in school. With this belief in mind, they only needed a short leap to the conclusion that more boys than girls are born with capabilities that push them to the top in math. Nonetheless, their leap is not based on a clean assessment of inborn abilities, but of the boys' and girls' performance after years of schooling, during which cultural expectations and socialization pressure are in full swing.

Larry Summers latched on to this study when he asserted that boys are innately predisposed to be at the more extreme end of mathematical talent than girls. But he and other folks ignored some other important facts when they push this innate-ability view.

Since that initial talent identification program in the 1970s, several million seventh and eighth graders have taken the SAT-M through annual talent searches. Interestingly, the 13-to-1 boy-girl imbalance reported in the early 1980s had dropped by the year 2005 to a difference of only 2.8 to 1.[8] This period of the drop coincides quite closely with the enforcement of Title IX legislation (the Equal Opportunity in Education Act), which, broadly speaking, is designed to ensure that both sexes have the same access and support for education-related activities. Although Title IX is most widely known for its impact on girls' and boys' opportunities in high school and collegiate sports, the original statute actually made no explicit mention of athletics. The goal of the legislation was to ensure equal opportunities for male and female students in all areas of academic life—including math and science education.[9]

Equitable access to math education seems to be a key source of reducing sex imbalances in achievement. A case in point is provided by data from the American Mathematics Competitions, or AMC.[10] The AMC are a series of contests sponsored by the Mathematical Association of America, held annually at more than three thousand high schools across the United States. Students who perform well on an initial AMC test are invited to participate in the American Invitational Mathematics Examination. Students who perform well on this exam are invited to the highly prestigious U.S. Mathematical Olympiad.

As an initial part of the AMC, students are asked to complete twenty-five problems in seventy-five minutes. The problems increase in difficulty from the beginning of the test to the end and span topics such as algebra, probability, geometry, and trigonometry.

Here are a few examples from one of the 2007 tests, the AMC 12 (for twelfth grade or below):

1. A piece of cheese is located at (12, 10) in a coordinate plane. A mouse is at (4, −2) and is running up the line $y = -5x + 18$. At the point (a, b) the mouse starts getting farther from the cheese rather than closer to it. What is $a + b$?

   (A) 6 (B) 10 (C) 14 (D) 18 (E) 22

2. Let a, b, c, d, and e be distinct integers such that $(6 - a)(6 - b)(6 - c)(6 - d)(6 - e) = 45$. What is $a + b + c + d + e$?

   (A) 5 (B) 17 (C) 25 (D) 27 (E) 30

3. The set {3, 6, 9, 10} is augmented by a fifth element n, not equal to any of the other four. The median of the resulting set is equal to its mean. What is the sum of all possible values of n?

   (A) 7 (B) 9 (C) 19 (D) 24 (E) 26

4. How many three-digit numbers are composed of three distinct digits such that one digit is the average of the other two?

   (A) 96 (B) 104 (C) 112 (D) 120 (E) 256

Don't feel bad if you have difficulty solving these problems, because this AMC test is designed to be difficult so it can distinguish between students performing at the highest levels in math. To give you some idea of how performance on the AMC 12 tests translates to other tests you may be familiar with, most students who score in the 99th percentile on the SAT-M (780–800 points) get the first three questions above correct, but the last question is only answered correctly by 44 percent of these high scorers. These tests are specifically designed to capture high-level math ability. (The answers and detailed descriptions of how to work out the problems can be found at www.artofproblem solving.com/wiki/index.php/2007_AMC_12A_Problems.)

Perhaps most interesting, however, is where the top-scoring boys and girls come from. The boys who score at the top come from a variety of backgrounds, but the top-scoring girls are all clustered in a small set of elite schools. Indeed, if one looks specifically at the data from the International Mathematics Olympiads and the China Girls Math Olympiad (which U.S. students qualify for after high-level performance on the initial AMC test, stellar performance on the subsequent American Invitational Mathematics Examination, *and* after doing well at the U.S. Mathematical Olympiad), as many girls come from the top-twenty-scoring AMC schools as from all other high schools in the United States combined. Unless you believe that girls with the highest level of math ability choose to attend only a handful of schools, these data suggest that most girls aren't being given the chance to reach their full mathematics potential. In other words, only a handful of schools are giving girls the support they need to succeed.

When they are not given the opportunity and support to excel, girls are not represented at the highest levels of mathematics, and this difference in numbers in itself helps to perpetuate a stereotype about the genetic basis of math sex differences. We may have a vicious cycle on our hands. Simply being made aware of stereotypes about how you should perform as a member of a particular group—a girl aware of stereotypes about gender and math—can degrade your ability to perform your best on important tests. Limiting girls' opportunities may perpetuate gender stereotypes, which in turn may stunt girls' scores further, limiting future opportunities, and so on.

Before we explore this cycle in more detail, however, let's take a step back and really look at what people mean when they assert that there are differences between the sexes that are genetic in origin.

HARDWIRING?

When the term "innate ability" is used to talk about sex differences in thinking or reasoning skills, it is usually intended to mean that there are hardwired, genetically determined variations in boys' and girls' brains and that their brains are different in the way they're organized and connected. So the conclusion often reached from this premise is that the sexes differ in innate ability to do, say, math or science. Let's look at the scientific basis for some of these claims. With respect to math, for instance, neuroimaging studies have found that, when people perform arithmetic and number-based tasks, they engage the inferior part of the parietal lobe in the brain.[11]

Some researchers have argued that this brain region is generally larger in men and than in women (even when controlling for overall brain volume). And because it sits right next to and closely communicates with brain areas involved in spatial navigation, reasoning, and attention, researchers have also suggested that this may give boys a boost in thinking about math in spatial terms. Being good at visualizing math problems or mentally rotating objects in your head—that is, being good at blending math and space—comes in handy for performing many types of calculations and is especially useful in geometry or trigonometry.[12]

Of course, before you take such suggestions too seriously, you should know that for every study that has found significant differences between the sexes in brain size and functioning, other studies do not support or verify these findings. Still other studies find evidence that contradicts them. One reason this is the case is that sex-difference studies often involve very small groups of people (sometimes fewer than several dozen participants), so results vary from one study to the next. Moreover, it is difficult to directly associate brain functioning with complex math performance because research on the brain systems involved in even basic mathematical processes is still in the early stages. Most of the math work so far has focused on how we understand number and quantities (such as twelve dots) and sex differences are not typically found for these basic numerical activities.

Some scientists have suggested that exposure to masculinizing sex hormones, particularly androgens, puts boys at an advantage in math and spatial skills, because androgens alter developing brain structure and function in a way that supports these activities. One way the androgen hypothesis has been tested is to study girls with congenital adrenal hyperplasia (CAH), a disorder involving prenatal exposure to excess androgens. Girls born with CAH often show play and social behavior more typical of little boys than girls and some studies have shown an advantage for girls with CAH on spatial-related tasks. But keep in mind that parents usually know about this condition from a very early age; any differences that are found could be explained just as easily by differential treatment by parents and other adults who know about the in-utero androgen exposure as by masculinizing sex hormones changing girls' brains to make them better equipped for math and science. Furthermore, other studies have shown no differences between CAH girls and non-CAH girls at all.[13]

Nonetheless, if girls' and boys' brains are organized differently *and* this is the reason why boys have an edge in terms of top scores on tests like the SAT-M, then it becomes hard to explain why a 13:1 boy-girl ratio has dropped so drastically in less than a quarter of a century. Brain structure just doesn't evolve that fast. A more likely explanation, and one put forth by one of the original talent study investigators, the late Julian Stanley himself, is that the ratio change reflects the fact that girls are getting exposed to math at earlier ages.[14] Compared to twenty-five years ago, boys' and girls' math education is more uniform, and today young girls are encouraged to pursue their interests in math and science more than they were in previous generations. In contrast to a quarter century ago, girls now have more opportunities to acquire the math tools they need to succeed. As a result, their ability to score at the top on tests like the SAT-M has skyrocketed.

Of course, a 2.8:1 boy-girl ratio still means that boys are outnumbering girls by almost 3 to 1 at the upper end of the SAT-M. But before you put too much stock in this imbalance, let's take a step back for a moment and acknowledge an important point. A ratio of 2.8 boys to every 1 girl is only meaningful in drawing conclusions about math differences *if* tests like the SAT-M equally estimate the abilities of talented boys and girls. If, for some reason, such tests don't capture math ability equally across the sexes, then any conclusions you might draw from this test about who has more or less talent will be flawed. This would be true whether we were estimating the boy-girl imbalance at 13 to 1 or 2.8 to 1. As it happens, there is some pretty convincing evidence that tests like the SAT-M don't capture ability equally across the board.

This is an often neglected finding. The SAT-M underpredicts girls' college math performance relative to boys'.[15] Simply put, when the SAT-M scores of boys and girls are matched, girls go on to earn higher grades in university-level math classes than their male counterparts. That's right, the same exact test score by a high school senior translates into a lowballing of girls' college math grades and an overestimation of boys'.

Why would this be the case? One reason has to do with the strategies that mathematically talented girls and boys tend to favor when solving the types of math problems seen on the SAT-M.[16] Girls are more likely to work all the way through a math problem using the exact procedures they learned in school. As a result, girls tend to outperform boys on problems in which following a specific step-by-step solution recipe is the most likely path to success. In contrast, boys outperform girls on unconventional problems that require unusual solution strategies. Boys tend to be more comfortable taking shortcuts than girls, and on tests like the SAT, where a large number of items need to be completed in a short amount of time, being able to apply your knowledge in quick and unusual ways has some advantages.

As an example, take the problem below, which appeared on a 1998 SAT-M.

A blend of coffee is made by mixing Colombian at $8 a pound with espresso coffee at $3 a pound. If the blend is worth $5 a pound, how many pounds of the Colombian coffee are needed to make 50 pounds of blend?

(A) 20 (B) 25 (C) 30 (D) 35 (E) 40

The answer: 20.

Solving this problem with a standard algebraic formula learned in school is fairly difficult. Not only that, it is time-consuming and leaves you open to simple mathematical mistakes. However, you can take a shortcut to solve this problem. Logic tells us that over half of the blend must be the cheaper espresso coffee because the price per pound ($5) is less than half the sum of the price per pound for each type: ($8 + $3)/2 would be $5.5 and the coffee is only $5 per pound. With this information, we know that only one possible answer option is correct because all of the others are greater or equal to half of the 50 pounds (that is, 25).

Boys are more likely to use this sort of logical shortcut than girls—a strategy that is obviously advantageous on timed tests.

Boys' tendency to rely on more flexible problem-solving approaches doesn't just occur at the high school level; it happens as early as elementary school. Despite the fact that there are not usually sex differences in mathematical achievement in the early grades, observation of classroom behavior shows that girls are more likely to use standard calculation procedures for arithmetic whereas boys often take more flexible, unconventional problem-solving approaches. For example, when asked to calculate 38 + 26, girls are more likely, in a step-by-step fashion, to first add the ones digits, 8 and 6, get 14, carry the 1, add that to the tens digits 3 and 2, and end up at 64. Boys, on the other hand, might decide that 30 and 20 is 50, and 8 makes 58; then 6 more is 64.[17]

Where do these strategy differences come from? In another study, when third- and fourth-grade students were interviewed about how they could solve arithmetic problems like those above, everyone reported that they knew about both the standard and unconventional problem-solving methods, but only the boys actually used the unconventional approaches.[18] The students mostly learned the conventional strategies in school (from largely female elementary school teachers) and the unconventional methods at home working with their brothers, uncles, or fathers on tasks involving building and measuring. At the elementary school age, children tend to pick up on behaviors and attitudes from same-sex adults. Because children are most likely to model behaviors that are specific to their sex, this may put boys at an advantage over girls in developing a diverse repertoire of problem-solving approaches.

The very tests used to gauge ability in math and science don't always assess boys' and girls' talents equally. Tests like the SAT—which have been used to assess students' talent for years—are better at predicting the performance of some students than others. Over the past several decades, there has been rising concern about standardized tests' accuracy in gauging students' knowledge and skill. This concern has become so great recently that the chancellor of the entire University of California system, Dr. Richard Atkinson, boldly suggested in 2001 the SAT in its current form be dropped as an admissions requirement for the University of California.[19] Atkinson and others have argued that relying too heavily on standardized tests like the SAT in college admissions decisions is dangerous because these tests don't provide every student with an equal opportunity to showcase his or her knowledge and skill.

I attended a University of California basketball game several years ago when they were playing Stanford University. The California-Stanford matchup is one of the biggest rivalries in college sports, so the fans for each team were doing their very best to distract their opponents. Bay Area basketball phenom and California freshman Jason Kidd was leading the UC Berkeley team that year and, although Kidd was a star on the court, it was widely known that it had taken him several attempts to score the minimum SAT score needed to matriculate. As Kidd stepped up to the free-throw line, members of the Stanford band held up a sign that said, "Hey Kidd. How do you spell S-A-T?" Kidd missed the shot.

S-A-T might have been three of the scariest letters in the alphabet for Kidd when he was trying to move to the university level. Even students who score in the top percentile worry about the test for years beforehand because of the importance of high scores for most admissions decisions. This is something Atkinson commented on when he recommended getting rid of the test for university admissions decisions. As Atkinson put it, "I have come to believe that America's overemphasis on standardized tests in general, and the SAT-I in particular, is compromising our educational system." Poor performance on tests like the SAT occur for many reasons, some of which are relatively divorced from a student's actual ability and skill.

THE THREAT OF STEREOTYPES

Think about a female high school senior (let's call her Taylor) who is not only mathematically talented, but has expressed interest in pursuing math as a college major. As we have just seen, because she is a girl, Taylor is more likely to favor math problem-solving strategies that are not always advantageous on timed tests. She also faces other hardships when taking important tests like the SAT-M, including unique pressures to perform well because she is female—pressures that her male fellow students don't face.

Consider what happens if, in the middle of taking the SAT-M, Taylor thinks about the well-known and widely held stereotype that "girls are bad at math." She is at risk for being judged by the stereotype, and research shows that just being stereotyped negatively is enough to drive down performance. Interestingly, poor performance in the face of negative stereotypes—known as *stereotype threat*—is most dramatic for those girls who are the most interested in excelling at what they are being tested on. Think about this for a second. You might expect that girls with the highest levels of math ability would readily disprove a negative stereotype like "girls can't do math," but this doesn't seem to be the case. Rather, high-achieving girls are precisely the ones who, when faced with a negative stereotype about how they *should* perform, worry about confirming it. As a result, their scores suffer.

> Just being stereotyped negatively is enough to drive down performance

One of the best examples of how negative sex stereotypes can derail test scores comes from a study conducted in the late 1990s at the University of Michigan.[20] Researchers handpicked male and female college students who scored in the top 10–15 per-

> *Stereotype threat* is most dramatic for those girls who are the most skilled and most interested in excelling at what they are being tested on.

cent of all students in the country on the SAT or ACT (American College Test) they took before matriculating to college and asked these students to complete a difficult math test. The test involved advanced calculus and abstract algebra and was challenging to even the most mathematically proficient students.

Just as in Camilla Benbow and Julian Stanley's original math talent identification study, the researchers found sex differences in math performance. But this was not the entire story. Male students performed better than their female counterparts when, prior to taking the test, all students were told that the test they were about to complete had shown differences between the sexes in terms of past scores. When students were instead told that the test was gender-neutral—meaning that men and women were equally likely to score well—there were no differences in performance. Merely highlighting the possibility of a boy-girl imbalance was enough to negatively affect female students' scores—and these were female students at a top public university who, as demonstrated by their past achievement, already had the tools to score well in math.

Interestingly, similar experiments have been conducted with female students who scored on the lower end on the SAT-M. This new group of students was not high achieving in math and, frankly, was not really bothered by this fact, so their scores didn't suffer when the stereotype that "girls can't do math" was brought to their attention. Telling women who are neither that talented nor that interested in math that other women who took the test they are about to take scored low doesn't have much of an effect—perhaps because these women didn't really care about confirming the stereotype in the first place. In contrast, for female students with the ability to succeed and an interest in rising to the top, highlighting expectations of poor performance are quite threatening—hence the name *stereotype threat*.[21]

When highly capable women are made aware of how they *should* perform, they recruit additional working-memory and emotional centers in the brain to deal with this information.[22] These brain centers likely come into play to combat the negative thoughts and worries that arise from the idea that "girls can't do math" and, importantly, these same emotion-related brain areas are not as active when sex differences in math are not brought to the forefront of these women's awareness. When brainpower that could otherwise be devoted to math is instead redirected to controlling worrying, the test taker has fewer resources to support her math problem solving and, as a result, her performance suffers.

Relying on high-stakes tests like the SAT, particularly as a basis for the type of arguments that Summers made at NBER, ignores the questions of what skills or aptitudes such tests actually measure and whether their measurements are equally accurate for all test takers. The answers to these questions—which are still unknown after almost six decades—would provide a more balanced, realistic view of the division between the sexes' achievements in math and science.

EARLIER IN LIFE

Most psychology and brain researchers would have little problem with some of the views echoed in Summers's talk, such as the idea that human math and science reasoning stems from an innate ability to represent objects, space, and number, which manifests itself early on in life. The question is, do these abilities come online with more force in boys than girls?

One way to test Summers's remarks is to find out whether differences in boys' and girls' abilities that are important for math and science, such as spatial abilities, are ubiquitous in nature. In other words, is there scientific evidence to support the idea that, anywhere you look, girls and boys have different intrinsic aptitudes for math and science? A few years ago, Susan Levine, a colleague of mine at the University of Chicago, spearheaded an effort to do just this.

Susan and her research team had more than five hundred second and third-grade boys and girls from the Chicago area complete a number of different spatial tasks.[23] One of the tasks, the square rotation task, is a nice example of what the kids were asked to do. Children were told to select the figure that could be put together with the target figure to make a square. As you can see, to do this task you must rotate each of the figures in your mind's eye to see if it does indeed make a square when combined with the target.

The ability to rotate objects in our heads is important for success in math and science and is a crucial aspect of tasks we do every day. Just think about how you have to use mental rotation when you are lost in a new city and trying to figure out where you are located relative to a particular street on a map, or whether your car will fit in a small space where you want to try to parallel-park it. Levine and her colleagues were out to see if there were differences between the sexes in this ability. But the researchers went a step further and did something that at first glance might seem a bit odd. They looked at the family income levels, or what is often referred to as socioeconomic status (SES), of the girls and boys they tested. Levine's logic was that if there are differences in spatial ability that are innate to boys and girls, then boys should always outperform girls on spatial tasks. But if boys only out-

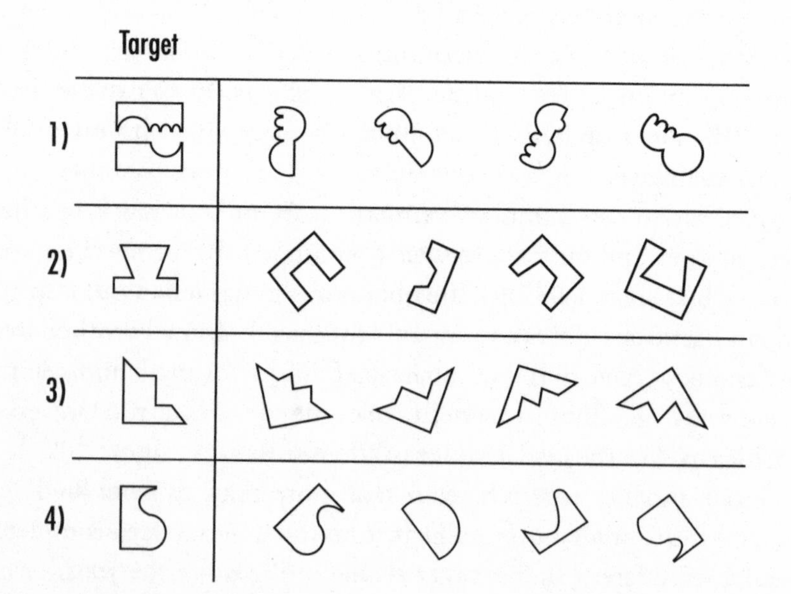

Square rotation task[24]

pace girls at some family income levels but not others, then it becomes hard to make an innate argument. After all, if differences in spatial ability are determined at birth, they shouldn't be related to how much money a child's family earns each year.

The researchers grouped the kids they tested into low-, middle-, and high-income levels based on their yearly family income. In 2000, the median household income for a family of four in Illinois was $46,064. The middle-income group in the study consisted of kids whose families brought in just about this much money, $39,373–$50,733 a year. The family income level of the kids who fell into the highest SES group ranged from $59,124–$124,855. In striking contrast, the range of incomes for the low SES group was $19,371–$26,242. Importantly, Levine made sure that boys and girls were about equally represented at each of the income levels.

When averaging across all family income levels, the boys did indeed perform better than the girls. But when Levine looked a bit more closely at her data, she found that this boy-girl gap in spatial ability only occurred for those kids whose families brought in the *most* money on an annual basis. Both the boys and girls from the poorest families performed equally badly.

What can we make of these results? One possibility is that kids from low-income families performed so poorly on the spatial tasks that there was no way to detect any differences across boys and girls in the first place. Maybe everyone in the low-socioeconomic-status group, regardless of his or her sex, bombed the test, making it impossible to find sex differences that might exist. But this poor-performance idea doesn't explain the findings because Susan went back into her data and specifically looked at the lower SES kids when they were performing at higher spatial abilities. To do this, she compared the spatial performance of the low-SES kids when they were in the spring of third grade to the performance of the middle- and higher-SES kids when they were in the fall of second grade. Although these kids were almost a year apart in school, this year helped to put everyone on equal footing with respect to their spatial abilities. Despite everyone generally being around the same ability level, Susan still found the same pattern of results: there was still a boy-girl imbalance in the high-income and middle-income families that wasn't there for the lowest family income level.

Although the researchers are still working to uncover all the possible reasons for their findings, they think they might have a clue—and it stems from boys' and girls' opportunities for spatial play. As it happens, activities like playing with Legos, putting puzzles together, exploring one's surroundings, and even playing video games can help kids develop spatial skills, but there are differences between the sexes and the classes in who engages in this type of play.

Take Legos as an example. A majority of the Legos sold in the United States each year are intended for boys,[25] but because Legos are expensive, neither lower-class boys nor girls see much of them. So boys from middle- and upper-class families get to hone their spatial skills through playing with Legos in a way that girls and kids from lower-class families do not.

Young boys are also usually allowed to navigate farther from home than girls, yet a kid has much more freedom to explore his surroundings and develop spatial skills when he lives in a middle- or upper-class neighborhood where it is safe to wander away. In short, there is ample evidence that the differences in boys' and girls' spatial abilities—at least by early elementary school—are largely dictated by the experiences these kids have. And it just so happens that wealthier environments offer more opportunities for boys and girls to go down different paths.

VERY EARLY IN LIFE

Showing that experience shapes spatial ability by elementary school is informative. But keep in mind that studies like Susan Levine's still don't tell us whether innate differences in spatial ability—or math- and science-related abilities more generally—are in effect very early on in life *before* socialization has had a chance to take over. This is a hard issue to deal with because the environment comes into play the second a child enters into this world.

One way to get at whether there are innate sex differences in math and science aptitude is to take a cue from some of our monkey relatives. Researchers at the Yerkes National Primate Research Center in Atlanta have found that, when given a choice, male rhesus macaques are more predisposed to play with wheeled and mechanical toys while female monkeys play with both wheeled and plush, doll-like toys an equal amount of time.[26] If we assume that male and female monkeys have not been socialized differently, the way young boys and girls might be, and indeed there is work suggesting this is true,[27] then it becomes difficult to explain differences in male and female monkey preferences by their environment alone. The Yerkes researchers think that differences in prenatal exposure to masculinizing androgen hormones are behind male and female monkeys' innate preferences for toys that serve different functions—say, toys that are mechanical versus not. You might think that male baby monkeys see other males playing this way and simply "ape" them, but if this were the case, the difference would still have to start somewhere.

Some researchers think these same prenatal hormonal effects extend to human children as well. It's been suggested that newborn baby boys are often more focused on objects (such as those mobiles that hang above a baby's crib) than are newborn girls.[28] These newborns have presumably not received differential treatment. If baby boys show a preference for objects that baby girls do not, then maybe differences in mechanical interest and abilities between the sexes do have a genetic basis.

Of course, if it is really true that baby boys have an innate interest in objects and their mechanical properties and that baby girls do not, then you would expect older male infants to have more spatial knowledge of objects and to be better at manipulating these objects than girls. After all, boys' innate object predisposition should drive them to interact more with those objects. But this is almost never the case.[29] Instead, newborn girls do show interest in objects and female, like male, infants display spatial knowledge of objects. Boys and girls largely follow the same developmental path.

In sum, from research with infants all the way through high school students, it seems that boys and girls share capacities that allow both sexes to develop talent for math and science. Of course, biology is an important determinant of cognitive ability. As we saw in chapter 3, genetics can explain some of the differences in cognitive functioning across people, but there is little evidence to suggest that these innate differences are strongly rooted in their sex. Nor can it be substantiated that biological differences are the root cause of differences in the numbers of men and women in math, science, and engineering.

IT DOESN'T HAPPEN EVERYWHERE

Regardless of what side of the debate you come down on concerning whether there are innate differences in girls' and boys' predispositions to develop spatial skills or attend to all things mechanical, one piece of information is hard to ignore: the math and science testing differences that Larry Summers talked about are not universal in nature. If these differences between the sexes don't occur everywhere, can innate influences really be that strong?

Recently, economists from the University of Chicago and Northwestern University got their hands on data from a 2003 Program for International Student Assessment (PISA) in which more than 276,000 fifteen year olds from forty different countries took identical math tests.[30] They discovered a strong relation between a country's attitudes toward women and the differences between male and female performances in math. The more emancipated a country was in its views about women's equality with men and its opportunities for women, the less the boys and girls differed in math test performance.

In countries such as Turkey, for example, where there is a huge gap in the cultural attitudes held toward men and women, there were marked sex differences in math performance: boys outscore girls. But in countries of sexual equality, like Norway and Sweden, this gap essentially disappears. In the United States sexual inequality is alive and well and, just as you would predict from other countries with these attitudes, the math gap between boys and girls on the PISA test is evident. You might have expected this from a society that, as recently as 1992, produced a doll—the Teen Talk Barbie—that, when its cord was pulled, would say things like "Will we ever have enough clothes?" and "Math class is tough!"[31]

There are strong relations between a country's attitudes toward women and the differences between male and female performances in math. The more emancipated a country is in its views about women's equality with men and its opportunities for women, the less the boys and girls differ in math test performance.

Given that the United States tops the list in terms of economic development of the countries at which the PISA study looked, these results suggest that a country's economic achievement alone doesn't lead to equal math proficiency between the sexes. The elimination of the boy-girl math test gap appears to stem from the improvement of the role of women in society. The more emancipated women are, the better their society's attitudes toward women and education, achievement, and success, and the less likely girls are to fall behind boys in math.

In many of the countries where differences between girls and boys in math were *not* apparent—for example, Iceland—there were also more girls than boys scoring above the 99th percentile in math achievement. It's hard to hold tightly to Larry Summers's view that men are more variable in their abilities than women (and thus more likely to score at the top) when the number of men at the highest levels is dependent on cultural and ethnic boundaries.

Studies like those conducted on the PISA data make it difficult to argue that genetic factors play a major role in the differential rep-

> If genetic factors played a major role in the differential representation of men and women in math and science, sex differences in math ability would exist in all countries—but they don't.

resentation of men and women in math and science. If this were the case, then sex differences in math ability should exist in all countries, but they don't. Rather, boys outscore girls in math in countries where men's opportunity for education and advancement outscores women's and where the attitudes put forth regarding men's and women's success in math are not uniform. The gender divide is a lot more complicated than Summers let on.

WHAT'S IN A NAME?

Taylor, a senior at Marin High, always took whatever math and science classes were available to her. She liked the material and found the classes interesting in their challenges. One of the classes Taylor liked best was her AP calculus class, in which only the most mathematically talented students in her high school enrolled. Aware that there were more boys than girls in her class, Taylor had never really given the cause of this difference much thought. As we have already seen, innate sex differences are not a likely cause of the makeup of Taylor's calculus class and don't explain the broader divide between men and women in math and science careers. So, what does? We have already touched on differential opportunities for boys and girls in spatial play and math education and there are even more subtle factors than these that affect the career imbalance in math and the sciences.

One factor is as simple as a girl's name. Economist David Figlio has shown that the more feminine a girl's name, the less likely she is to take calculus in high school.[32] Figlio argues that girls with more feminine names (think Isabella and Anna) associate themselves more with feminine ideals and are treated systematically differently by parents, teachers, and peers than girls with less feminine names (think Taylor or Madison). The end result is that girls with more feminine names are attracted to traditionally female coursework, such as the humanities and foreign languages, and shy away from disciplines like math and science.

Of course, the idea that something as simple as a girl's name may contribute to the gender divide is a bold claim, so when Figlio made it he had to be sure he had the goods to back it up. He does. For instance, you might be thinking that one obvious explanation for why a girl's name might be related to her tendency to take classes like calculus and physics has to do with her upbringing. Maybe girls given more feminine names are just reared in households where parents are more likely to create an environment that falls along traditional sexual identity and role lines and this is why these girls shy away from traditionally male-dominated disciplines. But family influence can't actually explain Figlio's data because he specifically looked at how the names of high-achieving *sisters* related to their likelihood of selecting into advanced high school math and science classes. Fortunately, parents often give sisters very different names—at least when it comes to their names' femininity—which offered Figlio a nice opportunity to test his hypothesis that a name can influence a girl's academic career even when the household environment in which she is raised is taken into account.

Figlio looked at all the courses taken by high school students enrolled in a large Florida school district between 1995 and 2001. He then investigated whether a girl's name (and specifically the femininity of her name) had something to do with her selection into calculus and physics classes—the math and science courses that are usually taken by only the most high-achieving students. Figlio found that girls with more feminine names were indeed less likely to take advanced calculus and physics than their less feminine-named sisters.

In order to conduct this type of analysis, David Figlio had to figure out what made a girl's name more or less feminine. This is no easy task—especially if you want to rate femininity without bringing in your own subjectivity or particular cultural connotations of what makes a name girly. But Figlio found a way.

In a nutshell, he separated all the most popular girls' names in the United States into their phonemes.[33] Phonemes are the particular sound combinations that make up words (for example the "a" in "Isabella"). Using all births in Florida from 1989 through 1996, Figlio created a mathematical model that related how probable it was that a particular phoneme would be associated with the name holder being female. Some phonemes are just much more likely to occur in girls' names than in boys names—like the "a" in "Isabella." The higher the number of phonemes in a particular name that were indicative of the name holder being female, the more feminine Figlio counted that name to be. On the highly feminine side you have names like Kayla and Isabella. At the other end, Taylor, Madison, and Alexis.

According to Figlio, then, one factor that contributes to why many girls shy away from math and science is as simple as their first name. Once you acknowledge that a girl's name can affect her academic path, it is only a small step to look for other subtle contributions to the differences between the sexes in math and science achievement. Indeed, merely being aware of the number of boys and girls in a particular environment can actually affect whether a female student will want to put herself in that situation. The bigger the ratio of boys to girls, the less likely a girl will opt to participate. This is true even if a girl has interest in the activity in question.

When Stanford University math and science majors, for example, were asked to watch a video about a science leadership conference that their university was supposedly hosting over the summer, and were then asked if they wanted to attend, female students were less likely to be interested if the advertising video depicted an obvious majority of men over women—in spite of the fact that the conference was centered around their major. Yet when the video portrayed an equal number of male and female conference attendees, the Stanford women wanted to go. The male-female makeup of the conference had no effect on whether male undergraduates wanted to attend. It was only for those people (women) already aware of an imbalance in their major for whom the demographics of the advertising videos had an impact.[34]

According to Claude Steele, one of the psychologists at Stanford who conducted the science conference study, people tacitly assess their prospects for success in academic arenas like math and science and people's interests follow these assessments. Our interest in an area of study increases when the prospects seem favorable, when we see other people like ourselves succeeding. When there is less evidence that we can succeed (such as the hurdles women face in high-level math and science), interest and willingness to participate decrease.[35] Think about this for a second. The end result is that we have a strong, self-fulfilling prophecy at work. The imbalance of men versus women in science negatively affects the interest that women have in pursuing opportunities in these arenas. As a result, women are less likely to show up in strong numbers, reinforcing the imbalance, and the cycle continues. You might be interested to know that when psychologist Mary Murphy, another author on the science conference study, was at Stanford, she noticed that the building where the mathematics department is housed only had one female bathroom—and it was in the basement. When there are clues all around us that women are not represented in strong numbers in math and the sciences, fewer women pursue careers in these areas, and nothing changes.

WHAT ABOUT BOYS?

Thus far we have primarily focused on situations in which boys outperform girls in science and math. What about the other way around? For several decades, talking about the educational disadvantages that boys might face in school was thought to carry sexist overtones or was believed to be "anti-girl" in nature. If you focused on the schooling difficulties of boys, you might be accused of ignoring girls or you would at least be in danger of falling back on 1950s stereotypes that little boys are rambunctious and can't sit still in school while little girls are good at adhering to their lessons. But this attitude is changing. In recent years there has been a push to highlight the difficulties boys have in the classroom and a good deal of evidence has been brought to the forefront that boys do sometimes underperform girls—especially at the elementary school levels.

Take the Wilmette school district in the Chicago suburbs, which Peg Tyre talks about in her 2008 book, *The Trouble with Boys*. This district decided to look into the "reversed gender gap" when the new school superintendent, Dr. Glenn "Max" McGee, came into office and noticed that boys were not only more likely than girls to get failing grades, but that girls were actually outperforming boys at the younger grade levels—especially in subjects like reading and writing.[36] Wilmette has taken a number of steps to make its subjects more "user-friendly" to all students and to ensure that there is ample time outside the classroom for expending the energy rambunctious students might otherwise bring into the classroom. The teachers think it's working, but, despite the recognition that boys may lag behind girls in some instances, the fact of the matter is that as students advance in grade level, boys still generally outscore girls on high-stakes exams.

Simply put, boys score higher than girls on standardized tests in math and science from the end of secondary school through graduate school. Just take a look at the Advanced Placement (AP) testing statistics put out by the College Board each year. The AP exams are the tests taken by high school students as a way to earn college course credit before they have ever set foot on a university campus. More boys than girls take the AP math, physics, calculus, and chemistry tests, to name a few, and a higher percentage of boys have clear passing scores. In 2007, 46.5 percent of boys taking the AP calculus test scored above the mean grade—a 4 or higher on the 5-point grading scale—compared to just 37.4 percent of female test takers.[37] The disparity between girls' and boys' scores on the AP tests carries huge implications for advancing in science and engineering fields. At the very least, if boys score higher than girls, they are at an advantage in gaining college course credit and selecting into higher-level classes when they matriculate. In the upper echelons of math and science, at least in countries where sexual equality has not yet come to fruition, there still seems to be a strong divide with boys outnumbering girls.

WHERE ARE WE NOW?

Although women receive more than half of all awarded undergraduate degrees, they receive far fewer in most STEM (science, technology, engineering, and math) disciplines and women comprise a minority of academic faculty positions in these areas—a proportion that decreases as faculty rank increases. These differences between the sexes are not just seen in the academic world—they extend to jobs outside the ivory tower as well.

Larry Summers's NBER comments prompted scientists and nonscientists alike to revisit issues of equal representation of men and women in math and science. Women are severely underrepresented at the highest levels in these areas. Although biology plays a role in people's development and success, a good deal of evidence suggests that (a) girls and boys come into this world with roughly equal abilities to develop into scientists and engineers and that (b) environmental subtleties—ranging from a child's name to his or her parents' income—can have a large effect on whether he or she will obtain the highest levels of ability and skill.

Merely making the type of comments that Larry Summers made during his NBER talk is enough to lower the performance of girls in important testing situations. As we have seen, if female college students are asked to complete a difficult math test and are reminded about differences between the sexes in math scores and achievement, their performance suffers relative to a male student's. Recent work out of my laboratory shows that this stereotyping effect is especially likely to occur in high-stakes tests where the pressure is on and everyone desperately wants to perform at his or her best. High-pressure situations magnify the environmental subtleties we have talked about, and not only contribute to the gender divide but to racial achievement gaps, as well.

Bringing up negative stereotypes about how your sex or racial group should perform—girls can't do math, blacks are not smart, even white men can't jump—is enough to send people into a spiral of self-doubt that uses up valuable brain resources that could otherwise work on the task at hand—resources that are already scarce in high-stakes situations. In short, the mere awareness of these stereotypes can lead to choking under pressure.

> The mere awareness of negative performance stereotypes can lead to choking under pressure.

In the next chapter we turn in more detail to how choking under pressure comes about on all-important tests—whether it is a female test taker, a minority student, or just someone wanting to perform at his or her best. We ask what conditions exacerbate less than optimal performance and explore how it comes to pass that some sail while others fail when the stakes are high and everything is riding on the next move.

CHAPTER FIVE

BOMBING THE TEST

WHY WE CHOKE UNDER PRESSURE IN THE CLASSROOM

Ninth-grader Jared never gave math much thought. A straight-A student in subjects ranging from English to history, he never worried much about his ability to perform at the top in school. Math was no exception, until he began to study for his preliminary SATs, or PSATs (students take the PSAT in order to practice for the actual SATs a few years down the road). A high score on the PSAT is a good omen for acing the real SATs. Stellar performance on the PSATs can also turn you into a "National Merit Finalist," which greatly enhances your chances of being admitted to the most elite universities.

Jared's parents had met while they were both undergraduates at Princeton University and had expected, since Jared's birth, that he would also attend their prestigious alma mater. However, even in his first year of high school, Jared was starting to worry about whether he could live up to this pressure. Jared's parents had recently suggested that he enroll in a test preparatory course for the PSAT to give him an extra boost from which every student can benefit in today's competition for top scores and schools. Jared had to admit that he was happy to have whatever practice he could get.

Roughly two months before the big PSAT testing day, Jared—who is African-American—arrived for his first test prep class. Jared found an empty seat toward the back, and sat down just as his instructor, a preppy guy in his late twenties, strolled in. The instructor had that look of academic cockiness that only guys with perfect educational pedigrees (from all-white boarding schools to the Ivy League) have, and introduced himself as a Ph.D. candidate in mathematics. He told Jared and the class that knowing the basics *and* being able to execute them quickly and flawlessly was the key to doing well on the math portion of the PSATs. Getting the simple parts of a problem out of the way allows you to devote more time and energy to the tricky parts lurking beneath the surface. Recognizing those tricky parts, Jared's instructor said with a snicker, separated those who belonged at state schools from those who would end up in the Ivy League.

The instructor wrote a problem on the whiteboard at the front of the room and began calling students—one by one—up to the board to solve the problems as quickly as they could. As Jared's name was called and he walked to the front of the class, he reminded himself that he was the master of the basics. "No sweat." Then he noticed his instructor looking at him, and his classmates—all of whom, incidentally were white—staring at him as well. Are these guys surprised to see a black guy in here? Jared thought, annoyed. Or are they just wondering if I know how to add and subtract?

Suddenly Jared's feelings of annoyance turned to panic and a whole new slew of thoughts crossed his mind: What if I mess up? What if I can't remember how to do even these simple problems? What if I look like an idiot in front of my math-whiz instructor and all these white guys? What if I don't do well on the test next month or on the real SATs next year? What if I don't get into Princeton? When his instructor rudely coughed—interrupting Jared's moment of panic—he realized he was standing at the whiteboard, pen in hand, with the problem "32 − 18 ÷ 3 = ?" in front of him. Instead of figuring out whether to subtract first or divide all he could think was, Oh, shit . . .

• • •

What is the first step in solving Jared's math problem? In one sense, the answer is quite simple. Standard order-of-operations laws in math say that division comes before subtraction so the first step is to divide 18 by 3. Next? Subtract this answer from 32. Fairly straightforward. But if Jared forgets about order of operations, and instead, because he is in a bit of a panic, automatically starts working on the problem from left to right because he is so used to reading in this direction, Jared will *get the problem wrong*. If Jared remembers to compute 18 ÷ 3 first, and even arrives at the answer 6 successfully, but because of his internal monologue of worries forgets that he is holding a 6 in mind and instead replaces it with 8, Jared will also *get the problem wrong*.

Such mistakes might, at first glance, appear surprising, given that this is a fairly simple problem for a successful ninth grader like Jared who has taken every math class available to him in school and has been flawlessly executing order-of-operations laws since he learned them. Yet, while computing the answer to "32 − 18 ÷ 3 = ?" may run smoothly when you are alone or under no pressure to succeed, you may react altogether differently when you are up in front of a bunch of people who expect you to fail.

How does solving a math problem as an entire class looks on affect a student's ability to come up with the correct answer? How might being in an important testing situation impede performance? What about when people are not in a testing situation at all, but find themselves struggling to calculate the tip on the dinner bill in the presence of their friends or colleagues? What is it about stares from your pals—especially well-educated ones—that can disrupt your ability to quickly compute 20 percent of $86? Or, what if Jared were not only enduring looks from his PSAT classmates, but also heard one say, "Maybe white men can't jump, but at least they can do math."

Although people may certainly be *motivated* to perform their best under stress, these environments can cause people to perform at their worst. The phrase *choking under pressure* has been used to describe what happens when people perform at a lower level than what they are capable of in high-stakes situations. But choking in performance can even occur in low-stakes situations, such as the practice test in which Jared now finds himself. Merely being aware of the fact that Jared's classmates might actually believe that there are racial differences in math ability is enough to cause Jared to screw up—even if Jared doesn't believe in these racial differences himself.

In the past several chapters, we uncovered some factors that influence successful performance in sports, academics, and business. In this chapter we take a closer look at the flip side of the coin—we ask why we sometimes fail to succeed when under pressure. Our goal is to unlock the secrets of *why* people "crash and burn" so that we can understand how to avoid failing in do-or-die environments. We will also learn something about *who* is most likely to choke under pressure—knowledge that will come in handy for predicting performance in the classroom, boardroom, playing field, and virtually all life pursuits.

MATH ANXIETY

So, why does Jared find himself frozen at the board? For years, researchers have studied why folks who are overly anxious about math perform poorly at it despite being competent on tasks in which math is not involved. Math-anxious individuals are overcome with feelings of fear when they are up at the chalkboard trying to complete a math problem or when they have to take a math test. People with math anxiety even dread sitting in a math class; just thinking about calculating a bill at a restaurant can send them into a panic.

But Jared doesn't fall into the "math-anxious" category. He likes math and excels at it. Nevertheless, he performs poorly for some of the same reasons that chronic math anxiety robs others of the ability to do math. This means that we can study people who have math anxiety and, from what we learn, figure out what it's like to be in Jared's shoes.

Until recently, most educational experts have assumed that math-anxious people perform badly because they have not acquired the skills to be good at math to begin with. People with math anxiety avoid math courses, learn less math when they are forced to take a math class, and stay away from math-related career paths. The end result is a lot of highly math-anxious people who don't know a lot of math.

But the view that math anxiety is just a proxy for poor math ability is changing, due in large part to the work of psychologist Mark Ashcraft, who found that one of the big reasons that math-anxious people bomb math tests is that the anxieties they experience while they are actually doing math diverts brainpower (such as working-memory) away from the math itself. When worries and self-doubt flood the brain, it's very hard for the brain—and you—to function properly. Of course, when you are performing an activity that is best run off on autopilot (like a forehand in tennis that you have hit perfectly thousands of times in the past) then distraction may not be such a bad thing—it's the overcontrol that results from worrying that can really mess you up in these automated athletic pursuits (more about this in the chapters to come). But when you juggle numbers and calculations in your head, and worries deplete the working-memory resources you need to concentrate effectively, your performance can be derailed.

Worry explains a lot of Jared's issues at the whiteboard as well. He's in a situation where he must demonstrate his competency and, rather than think about the math problem at hand, he is thinking that the people around him *expect* him to perform poorly because he is black. Just like the math-adverse, Jared's preoccupation with the situation is pushing him to fail.

Ashcraft started his career studying how people learn math, so now that his research is focused on understanding what goes wrong in people to give them high math anxiety, he is constantly being asked about what led to the shift in his interests. To answer, Mark tells a story about his daughter who was doing her elementary school math homework at the kitchen table one day. As a math researcher, he was obviously very interested in what she knew, so Mark asked her why she multiplied before she added. She responded angrily, "You just do it that way."

Apparently this is exactly what her teacher had said when one of her classmates had asked the same question. The teacher gave no explanation for order of operations; she got defensive because she likely didn't know the answer and was nervous about doing math. Indeed, as Mark has discovered in his work, elementary education majors—those who will go on to be elementary school teachers—have the highest levels of math anxiety of any college major in the United States.[1] This fact makes you wonder about the skills of some teachers and it has pushed Mark to investigate math anxiety.

Mark Ashcraft has focused a lot of his math anxiety work on university students. This is in large part because college is where career choices are made and where a student's intent to avoid math classes can send him or her down a specific career path. When a student chooses an early education major, by the way, he has opted for a curriculum that, generally, includes very little math. To figure out which students are more or less math-anxious, Ashcraft asks people to answer questions about how nervous they get in math-related situations like "being given a pop quiz" or "reading a cash register receipt." Those students who report they hate these situations are classified as highly math-anxious. Those students who report that these situations are "no sweat" are categorized as having low levels of math anxiety. The questions[2] are below:

How anxious would you feel about the following activities?

1. Receiving a math textbook.
2. Watching a teacher work an algebra problem on the blackboard.
3. Signing up for a math course.
4. Listening to another student explain a math formula.
5. Walking to math class.
6. Studying for a math test.
7. Taking the math section of a standardized test, like an achievement test.

8. Reading a cash register receipt after you buy something.
9. Taking an examination (quiz) in a math course.
10. Taking an examination (final) in a math course.
11. Being given a set of addition problems to solve on paper.
12. Being given a set of subtraction problems to solve on paper.
13. Being given a set of multiplication problems to solve on paper.
14. Being given a set of division problems to solve on paper.
15. Picking up your math textbook to begin working on a homework assignment.
16. Being given a homework assignment of many difficult math problems, which is due the next time the class meets.
17. Thinking about an upcoming math test one week before.
18. Thinking about an upcoming math test one day before.
19. Thinking about an upcoming math test one hour before.
20. Realizing that you have to take a certain number of math classes to meet the requirements for graduation.
21. Picking up a math textbook to begin a difficult reading assignment.
22. Receiving your final math grade on your report card.
23. Opening a math or statistics book and seeing a page full of problems.
24. Getting ready to study for a math test.
25. Being given a "pop" quiz in a math class.

You might think that people would be hesitant to admit that they are anxious about math, but they're not. As Ashcraft explains, it is socially acceptable in the United States to say you are bad at math. This attitude is very different from the attitudes that Americans have toward other academic subjects. You don't hear folks bragging that they can't read. Perhaps because it's easier to avoid math-related tasks than reading-related ones, people feel that it's okay to admit they don't like math, but our society's socially acceptable math phobia is a problem. The U.S. ranking in mathematical proficiency relative to other countries is low.[3] Of course, the fact that people are willing to divulge their math anxieties did make it easy for Ashcraft to study people who are math-adverse.

In one study,[4] Ashcraft began by having college students do simple addition problems in their head like "7 + 9 = ?" or "16 + 8 = ?" Pretty easy. But then he made things a bit more interesting. Ashcraft asked the students to do some more problems, but this time the students were given six random letters (such as "BLFMCX") to hold in memory while they figured out the math answers. After solving the math problems, the students repeated back the letters they were holding in mind to an experimenter sitting next to them.

Everyone was pretty good at doing the math task on its own. It was fairly simple addition and these were college students, after all. But when they had to perform both the addition task and the letter memory task together, the math performances were not as good. Doing two things at once is usually harder than doing one thing at a time, so Ashcraft's results are not surprising, but the students who were highest in math anxiety showed the most math errors when they solved the addition problems while also trying to remember the letters. Sure, students low in math anxiety did a little worse on the math when they had to perform the math and letter memory task at the same time, but their slip in performance was nowhere as large as what happened to the math-adverse students. To put it simply, math phobics—when faced with both the addition problems and the letter memory task—collapsed.

As it happens, this secondary task wouldn't have been such a problem if people were trying to learn a new language rather than perform demanding calculations in their head. In fact, in language learning, distraction can help. As discussed in chapter 2, for tasks that operate best outside of working-memory, such as soaking up a new language or hitting an easy putt in golf, distraction can help us steer clear of the specifics (like only learning particular word combinations during second-language learning or focusing too much on the step-by-step unfolding of your golf stroke). Yet when people are trying to juggle multiple things in mind at once—say a math problem, worries about math, and several random letters—something goes awry. And it starts with the math.

Back to the math study: Because everyone, regardless of math anxiety, could perform the simple addition flawlessly when there was no letter task involved, these results *can't* merely be explained by the premise that math-anxious individuals know less math. Instead, what Ashcraft thinks is happening is that, when math-anxious people do math, they worry—about the math, about how they will perform, about how they might look to others. These thoughts capture their working-memory so they don't have that much brainpower left to focus on the math itself.

When math-anxious people are only doing simple math, the worries that flood their brains aren't such a problem. People have enough working-memory or cognitive horsepower to cover worrying as well as the simple mathematical computations. But when math-anxious individuals have to perform a simple math task and also hold letters in memory, there is no room left for worries. Something gets sacrificed and unfortunately it's not the worries but the math.

Even though Jared is not generally math-anxious, when he gets to the whiteboard in his PSAT class, he is worrying just like someone who is math-adverse. And although Jared doesn't have to hold letters in memory while doing math, the problem he is faced with ("32 − 18 ÷ 3 = ?") necessitates that he keep several things in mind at once. So, when the worries come, Jared's mathematical ability suffers.

UNPACKING THE PREFRONTAL CORTEX

My Psychology Laboratory at the University of Chicago is set up much like the room Jared will encounter when he goes in to take his PSATs. There are tables with computers where students can sit alone or in a group and there is an experimenter, usually one of my graduate students, who serves as a proctor.

Lately, in my lab, we have been very interested in testing situations such as Jared's. My students and I want to know how people's awareness of a negative stereotype that others hold about them can result in poor performance—stereotypes such as "Girls can't do math," "Blacks aren't very smart," and even "White men can't jump." As we saw in chapter 4, the term *stereotype threat* has been coined to describe this phenomenon. Interestingly, when stereotyped in this way, people don't perform badly because of some inherent inferior ability, but because they are aware of how they *should* perform. Even more dangerous is that the performers don't have to endorse the stereotype themselves; they just have to think that others believe in it.

In one of the first studies exploring stereotype threat, psychologists Claude Steele and Joshua Aronson asked high-achieving African-American and white students at Stanford University to complete a portion of the Graduate Record Exam (GRE).[5] Prior to doing so, some students were asked a number of background questions—including what race they were. Other students were not asked about their race at all.

The researchers didn't find any difference in GRE performance between the white and black students when race wasn't reported. But when students did report their race before the test, African-Americans performed worse than whites. Having students identify their race made them think about the stereotype that "blacks are not as intelligent as whites." This idea is enough to bring down black students' performance in situations where intelligence is being gauged.

Merely being aware of a stereotype can bring down your performance

The damage that occurs when people are made aware of a negative stereotype about their ability is even more pernicious than merely causing poor performance across the board. People who are highly skilled and place a great deal of importance on their ability are hardest hit by negative assumptions about whether they can succeed. Just think about the black students from Stanford. You would expect that students at one of the nation's most prestigious schools would be more likely to disprove a negative stereotype about intelligence rather than reveal it. Because these high-achieving students don't want to perpetuate a stereotype and because they are conscious of it, these worries detract from their performance and these highly capable people suffer the most.

Indeed, it's not just people constantly aware of the prejudices held against their intellect—such as a black kid in a room full of white students or a girl about to take the math section of the SAT—whose performance suffers when negative expectations arise. White men do worse on the SAT when they are first reminded that Asians are good at math or when the quality of their educational pedigree is brought into question.[6] Let's take a step back for a second and look into the brain to figure out exactly how this debilitated performance arises.

Up to this point I have described working-memory as general-capacity horsepower—meaning that it supports our ability to work with information, regardless of what that information is. Another way to think about working-memory is as a flexible mental scratch pad. Working-memory helps you keep relevant information in mind (and irrelevant information out) when you perform a particular task. When worries flood the brain, whatever these worries may be, they deplete working-memory resources that would otherwise be available and your performance can suffer.

Working-memory is housed in the prefrontal cortex, which works with all different kinds of information, but certain parts of the prefrontal cortex are devoted to supporting particular types of information.[7] Tasks that are more verbal in nature, for example, tend to activate areas of the left prefrontal cortex. This is because doing something verbal, like remembering a phone number, involves talking in your head and in most adults the brain areas supporting language are largely constrained to the left side of the brain. In contrast, spatial tasks, like rotating an image in your mind's eye, are thought to occur

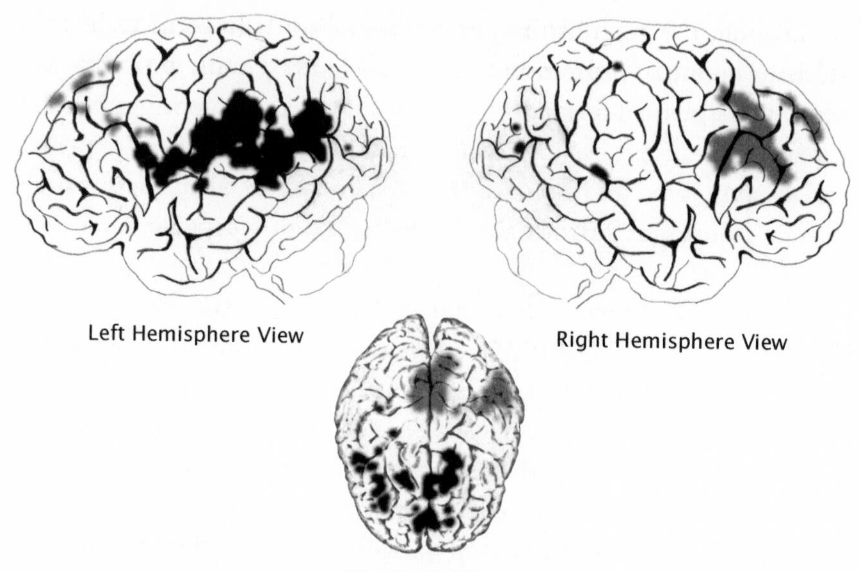

Brain regions most active during the performance of a verbal working-memory task (in dark shading; note largely left side activation) and brain regions most active during the performance of a visual-spatial working-memory task (in light shading; note largely right side activation).[8]

more in the right prefrontal cortex. This means that even though your cognitive horsepower can be thought of as quite general in nature, you can also carve it up into separate pools of mental resources devoted to working with particular types of information.

When pressure-filled situations create an inner monologue of worries in your head that taps verbal brainpower, performing activities that also rely heavily on these same verbal resources is more difficult. Doing two things at once that rely on similar brain regions is generally harder than doing two things that call on separate pools of brainpower, because in the first case there are just fewer neural resources to go around. This points to algebraic word problems as a good candidate to reveal the choke—at least in contrast to space-based math like geometry. Even more striking is that merely changing the way a math problem is presented can alter how much it taxes verbal brainpower and whether people will flub the problem under pressure.

Overall, math problems presented horizontally seem to be more reliant on verbal brain resources than the exact same problems presented in a vertical format.

Horizontal Problem:

32 - 17 =

Vertical Problem:

$$\begin{array}{r} 32 \\ \underline{-\,17} \end{array}$$

When people solve horizontal math problems, they tend to *verbally* maintain the intermediate steps in their head just as they would if they were reading a line of text from left to right. In contrast, people tend to solve vertical problems in a *spatial* mental workspace similar to how these problems are solved on paper. In the latter case, people actually imagine working out the problem in their head as they would if they had pencil in their hand. These visualization processes call on some of the same right prefrontal brain regions used to do spatial tasks like mental rotation. So merely changing how a math problem is presented on a page can change how your brain goes about solving it, which has big consequences for whether people will get the problem right or wrong when the stress is on.[9]

A few years ago, my students and I conducted a study in which we asked female college students to complete a series of subtraction and division math problems. Some of the problems we had the students solve were oriented vertically and some horizontally, although the problems were *exactly* the same. The only thing we changed was how they were presented on the computer screen. Before the female students completed the problems, we randomly pulled aside half the group and told them that at most schools, men outnumber women in math majors and majors with math as a prerequisite, and that there is a good deal of evidence that men consistently score higher than women on standardized tests of quantitative ability. We didn't mention differences between the sexes in performance at all to the other half of the female students in our study.

Sure enough, the female students reminded of differences between the sexes in math did worse than the women who did not receive this information. However, we saw the most evidence of poor performance on the math problems that were presented in a horizontal orientation. These were the problems that we knew would require a lot of verbal brainpower to solve. Performance on the exact same problems presented vertically, so that the students switched to more spatial parts of the brain to solve them, was not eroded under the stress of the stereotype that "women can't do math."

This pattern of poor performance is not limited to women and math. In a follow-up study, we asked both male and female college students to perform the same math problems we gave the female students above—written either horizontally or vertically.[10] This time, however, some of the students were simply asked to perform at their best while we pulled out all the stops for other students to get them to feel the stress of taking a high-stakes test. As I described in chapter 1, we turn up the stress in our lab in several ways. First, we offer students money for stellar performance. It's certainly not the same amount of money that would be given out as a college scholarship, but it's enough money to make college students anxious about not doing well on our test. We also tell students that other people are depending on them to succeed and we even videotape them while they do the math. This feeling of "all eyes on me" is similar to what students feel when they know that their test scores are going to be public knowledge.

Just as female students who were reminded about differences between the sexes' performance in math did not test well, those students who took our high-pressure test performed about 10 percent worse on the math problems than students in the low-pressure situation. Again, this poor performance was limited to the horizontally written math problems that tapped verbal brainpower the most. The exact same problems oriented vertically were performed equally well by people in the low-pressure and high-pressure tests.

In this particular study, we also asked the students whom we had put through our kitchen sink of stress what they thought about while they were doing the math. They told us they had worried—a lot. The pressured students had thoughts running through their head while they did the math like "Don't screw up!" "I hate math," "I'm not good at doing math in my head," and "Ahh . . . got one wrong!" The more that the test takers worried under pressure, the worse they performed. These findings provide pretty nice evidence that an internal monologue of worries is one of the big contributing factors to choking under pressure in academics. They also provide a possible clue for how to deal with stress. Merely rewriting a math problem in a format more conducive to spatial problem solving could take some of the load off verbal brain resources, which in turn might limit the damage that worries can do. As an example, rather than doing even fairly simple arithmetic in your head, rewriting 51 − 19 to be

$$\begin{array}{r} 51 \\ \underline{-19} \end{array}$$

and then doing the calculation in your test book might help you avoid silly mistakes that occur when worries and numbers compete for verbal brain resources. Also, by rewriting problems in this way, you are "cognitively outsourcing" some of the information you might have once mixed up in memory to the paper in front of you. By letting the paper serve as an external memory source—one that is relatively more free of worries than the prefrontal cortex—your performance is less likely to suffer when you are under pressure to succeed.

THE STRESS SPILLOVER

Worries are a big culprit in academic underperformance—worries that stem from a habitual anxiety about math, from not wanting to confirm a negative stereotype, or even from a fear of performing poorly when someone else is depending on you. These worries are even more problematic than my colleagues and I initially thought, because they don't immediately subside when the pressure-filled situation is finished.

As an example, a high school girl may not only stumble on the quantitative portion of the PSAT because she is thinking about the stereotype that "girls can't do math," but this preoccupation with her sexual identity can also translate into trouble when she starts the verbal section of the very same test. Logically, a math stereotype shouldn't apply here. If anything, girls might anticipate doing better on a verbal test since there is a general belief in our culture that women have superior verbal skills than men to begin with. But this isn't what happens. The brainpower compromised by worrying about girls and math doesn't immediately rebound when the math performance is over. As a result, any type of difficult thinking or reasoning you do after an initial bout of stress can suffer.

Just think about what might happen if a female high school student has her Algebra II class in the day's first period and then moves on to English in second period. Any worries or stress she feels as a result of being a girl doing math may inadvertently deplete the working-memory she is able to harness for analyzing literature in her English class.

Recent brain-imaging research tells us why this stress spillover occurs. When people are asked to do a difficult math task in front of an audience, a task that involves continuously subtracting 13 from a four-digit number that starts at say, 4381, heart rates increase, their palms sweat, and they report feeling anxious and stressed-out. This is exactly what neuroscientist Jiongjiong Wang and his colleagues at the University of Pennsylvania were after when they asked people to do this subtraction task while their brains were being scanned using fMRI.[11] Wang even urged the students to go faster and faster as they did the math task in order to stress them out so that he could see what happened in their brains when they were under a lot of pressure to perform well.

Under stress, several brain regions showed increased neural activity that paralleled people's reporting of how stressed-out they were, including the right prefrontal cortex, which is associated with the experience of negative emotions like sadness and fear and increased vigilance. Even more interesting, however, was how the brain fared when the stressor was over. The increased activity that occurred under pressure in the prefrontal cortex didn't immediately subside when the math task was finished. Even when people were just told to relax and remain calm after the stressful subtraction task was done, brain areas involved in negative emotions and heightened attention (such as the right ventral prefrontal cortex and the anterior cingulate cortex) were still doing an awful lot of work.

The right prefrontal cortex is an important player in the "flight" component of our "flight-or-fight" response to stress. Evolutionarily speaking, it's not so hard to imagine how a prolonged state of heightened vigilance would be a good thing—certainly beneficial if you need to flee a stressful situation and maintain awareness of your surroundings once you are out. But keep in mind that, when the body shunts extra blood to the right prefrontal cortex, activation in the opposing left prefrontal regions sometimes decreases—indeed, this happened when people were asked to count backward by faster and faster speeds in Wang's study. The end result is that brain areas predominantly involved in verbal thinking and reasoning are not fully on line, which can decrease your ability to do complicated math tasks that rely on this verbal ability. Unfortunately, because this shift doesn't die down right away, your cognitive horsepower may be generally stunted long after the pressure-filled situation that originally triggered your flight response is over.

DO YOU THRIVE OR DIVE UNDER PRESSURE?

In chapter 3 we learned that some people have more working-memory (aka cognitive horsepower) than others and so perform better on academic tasks such as problem solving, reasoning, and reading comprehension. Yet, despite the fact that those highest in working-memory are usually very well equipped to perform at the top, as we talked about briefly in the first chapter, these people are actually the most likely to fail under pressure.

I discovered this phenomenon when I was a graduate student at Michigan State University working with my Ph.D. adviser, Dr. Thomas Carr. Tom was nice enough to let me set up an exam room in his lab so that we could really roll up our sleeves and begin to figure out why some people but not others choke under pressure. We had divided roughly one hundred Michigan State undergraduate students into groups based on whether they scored at the top or bottom of several measures of working-memory. We then asked all of the students to perform a series of math problems—first in a low-pressure practice situation and then again in a high-pressure test.[12]

Not surprisingly, the high-powered students performed the best on the math problems under practice conditions. However, when taking the same math test in a high-pressure situation, the performance of those highest in cognitive horsepower fell to the level of those lowest. The performances of students with the least amount of working-memory actually didn't suffer under stress.

We were quite surprised by our findings. We had thought that the poorly performing students would really take a dive when the pressure was on. After all, they didn't have a lot of cognitive tools at their disposal to begin with, so we thought that they would be in trouble when pressure compromised their working-memory. Yet we found the exact opposite. Of course, our results did coincide with some of the work we have talked about thus far, such as the black Stanford undergraduates who, despite being at one of the top academic institutions in the country, choked when they had to indicate their race before they took a test. Or the female students at University of Michigan whose high math scores fell when they were reminded of the differences between the sexes in math competency.

What is going on? There are several reasons why higher-working-memory people are so susceptible to choking under pressure. The first has to do with how they approach difficult problems to begin with.

The best students usually rely on difficult problem-solving strategies (versus simpler shortcuts) to compute math problems like the ones that show up on the SAT or GRE. If you have the brainpower to successfully compute answers in complicated ways, you tend to use the resources at your disposal. In most situations, this extra horsepower is advantageous. Yet when you find yourself in a high-stress, timed-assessment situation and you are robbed of some of your normal computational power, your ability to perform at your best may be in jeopardy.

Strategy choices are not the only answer to why pressure-filled situations affect people higher on the working-memory continuum. Individuals with the most cognitive horsepower tend to be bad at downplaying the importance of high-pressure testing situations when they find themselves under the gun, so they also have a hard time easing the tension when the stress is on. High-performing people really feel the pressure, which hurts their ability to succeed.

A few years ago, a group of French psychologists demonstrated this high-powered disadvantage quite clearly.[13] They asked college students at the University of Provence to complete a difficult puzzle task that is frequently used as a metric of intelligence. In this puzzle task, called Raven's Progressive Matrices, people are presented with increasingly difficult patterns that contain one missing segment and are asked to choose which segment best completes the pattern from a number of available options. To be successful at Raven's, you must be able to reason about how all the pieces of the puzzle go together. This involves holding and manipulating a lot of information in—you guessed it—working-memory.

For some of the University of Provence students, the task was described as a measure of intelligence and reasoning ability associated with overall success in math and science. This certainly turned the pressure on, because those students who did poorly on the task would be considered unfit for a career in science. For the other half of the students, the puzzle task was simply described as measuring simple attention and perceptual capabilities. From the perspective of this

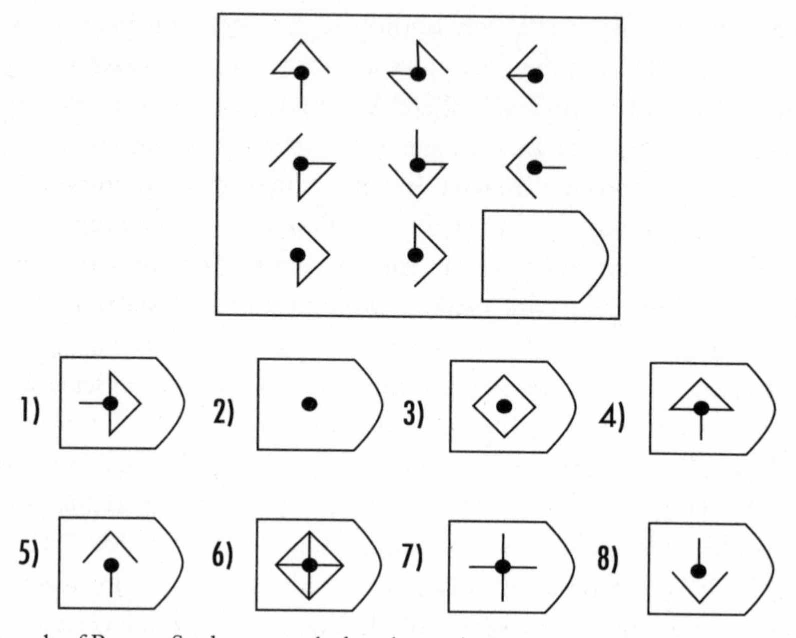

Example of Ravens: Students are asked to choose the missing piece among the proposed ones.[14]

latter group, the task didn't really assess much of anything so they had no reason to sweat their performance on it.

Not surprisingly, when students didn't think their puzzle performance was meaningful, the higher their working-memory, the better they performed. Those with more cognitive horsepower aced the test when the pressure was at a minimum. However, when the test was described as a measure of students' math and science aptitude—a pressure-filled situation—those highest in cognitive horsepower fell to the level of those lowest. The high-powered and low-powered students performed equally poorly.

After completing the puzzle task, all the students answered questions about their performance. If the students so chose, they could use these answers to excuse poor performance on the test they had just taken. People were asked to report, for example, how well they had slept the night before. If a student reported that he or she had not slept well, then this gave the student a nice excuse for bombing the test.

High-powered students were much less likely to provide an excuse for their performance. In a sense, these high-powered guys (and girls) were not good at finding ways to see a potentially stressful testing situation as not so pressure-filled. Although you might think that a "no-excuses" policy is always best, if you are able to take some of the pressure off yourself during an important test by reinterpreting the situation as something less stressful, less diagnostic of your ability, or less "do-or-die," you may be able to turn a potentially poor performance into a good one. Students lower in working-memory were quite good at this. They readily made excuses for themselves and, as a result, they could turn a stressful testing situation into something far less meaningful.

Less stress means less worrying and less likelihood that your performance will suffer when it counts the most. Less experience ruminating in general also translates into a lower likelihood that your brainpower will be compromised by worries when they do arise. This is because the thinking and reasoning abilities of chronic worriers are especially likely to be impaired when they do start to worry.[15] While you might imagine that people who worry all the time would be so accustomed to fretting that it would no longer be a big deal, the more you have a tendency to worry, the less working-memory you have at your disposal when you do start. It's as if chronic worriers are too good at worrying. They do it to such an extent they don't have brainpower left over for anything else.

This idea that stress can have different effects on different people depending on whether they are chronic worriers or depending on how they interpret a pressure-filled situation is something that one of my graduate students, Andrew, has been very interested in as of late. Andrew has been focusing on how people's physiological response to an intense situation—think clammy palms, increased heart rate, a sweaty brow—affects their test performance. Andrew wants to know how some people interpret these states as a cue to thrive, "My heart is racing, it must mean I am motivated," and how these bodily states are interpreted by others as a cue to dive, "Oh, shit, my heart is racing, I am really feeling the pressure now." Andrew thinks that getting a handle on this mind-body relation may be the key to understanding who will fail versus who will sail when the pressure is on and he has been having some remarkable success with this line of thinking.

As a jumping-off point, Andrew took a cue from a psychology experiment conducted in the 1970s on a swaying rope bridge that hangs 230 feet above the rocks and freezing water of British Columbia's Capilano River.[16] In the experiment, an attractive female accomplice intercepted men coming off the rope bridge and asked for their help in completing a short questionnaire. She also offered them her phone number in case the men wanted to hear more about the project she was conducting at a later time.

Although these types of surveys usually have an extremely low callback rate, an unusually large number of the men called the woman back. The men, having just walked across the high bridge, which left their hearts pounding and their palms sweaty, interpreted their physiological arousal as a cue about their attraction to the woman. Thus they were more likely than usual to pick up the phone and call her back. Some of the guys even got up the nerve to ask the female experimenter out for a date. When this experiment was also conducted with the same woman on a much lower bridge (where there was no opportunity for the men to be revved up by heights), the callback rate was at its usual low. So we know it was not the woman, but the bridge, that drove the men's responses.

My student Andrew wondered if the way people interpret their body's response to stressors might not just be limited to dating, but might also apply in high-stakes testing situations. So he asked students to take a fairly difficult math test. After they'd taken the test, Andrew had everyone spit into a tiny tube so that he could get a measure of each person's cortisol levels.[17] A hormone produced by the adrenal gland, cortisol is associated with several stress-related changes in the body, such as lower pain sensitivity and a quick burst of energy under duress. When people are in stressful situations, cortisol is secreted at higher levels and, because of this, it is often referred to as the "stress hormone." This means that cortisol is a quick-and-easy way to get a handle on a person's stress level at a particular time.

For some of the students, the higher their cortisol levels, the worse their math performance. Others, however, showed a very different pattern of results: the higher their cortisol levels, the better they performed! When Andrew took a closer look at what was driving this difference, he found something rather interesting: those people who did the worst when their cortisol was highest were also the ones who had, in a previous experimental session, professed to be extremely anxious about doing math. The people who performed better as their cortisol went up were the ones who were not at all math-anxious.

Of course, we already know that people who are overly anxious about math are likely to stress out when put in a math-testing situation. After all, highly math-anxious individuals are categorized as such because they get sweaty palms, a queasy stomach, and rocketing cortisol levels when they are faced with the prospect of doing math. Students who were low in math anxiety also had the same high cortisol levels and concomitant bodily reactions in the testing situation, but they just interpreted their physiological reaction differently. Interpreting the situation and your bodily response in a positive rather than a negative light may be a key to performing well when it counts the most.

This is good news, especially for a student about to do a problem at the board in class, and even for you when you find yourself in a highly stressful situation—say, when you are getting ready to give a big speech or negotiate an important business deal. If you can manage to interpret your body's response to the situation as positive, as a call to action, you are likely to thrive. But if you interpret your body's response as a sign that you are in a bad place with no way out, the worries and ruminations that result may send you into a "choke."

WHAT HIGH-STAKES TESTS TELL US

In March 1999, the California state legislature passed a bill requiring that all California public high schools administer an exit exam to their graduating students. The stated rationale behind the exit exam was to "ensure that pupils who graduate from high school can demonstrate grade level competency in reading, writing, and mathematics." The California exit exam is a two-part test involving mathematics and English language arts. Students must pass both parts to earn their diploma.[18]

California is not the only state to have implemented an exit exam policy in recent years. As of 2007, twenty-two states required exit exams and the number of states pushing the exam policy is increasing every year. Soon, over 70 percent of United States students will need to pass some sort of exam to earn a high school diploma.

On the surface, the exit exam seems quite logical—a positive move by the states to ensure that all students have a standard level of knowledge and competency when they graduate. But what if I told you that, as a result of California's exit exam policy, graduation rates in the state have declined by as much as 4.5 percent and that this decline is primarily concentrated among low-achieving minority and female students? Would you still think the exam is a good idea?

Some may try to account for this graduation-rate decline by assuming that women and minorities are less prepared to take the exams. If less able students fail at a higher rate, then the exit exam is doing its job. Right? But this logic doesn't work, because even if you only look at California students who all have similarly low scores on the array of state-mandated low-stakes standardized tests taken during high school—that is, students performing near the bottom of their class of every race and both sexes—being a minority or girl in this low-achieving group still means that you fail the exit exam at a higher rate than your white male counterparts. Having the pressure of graduation on the line in these exit exam tests seems to hit some harder than others.

Most students, regardless of their background, feel the pressure to pass the exit exam. After all, failure means not graduating. Yet minorities and women face an added burden. Not only do these students fear poor performance, they also walk around under the weight of negative stereotypes that impugne their intelligence. And, as we have seen throughout this book, such stereotype threats can affect whether students are able to show what they know. Using a high-stakes exit exam to determine who is eligible to graduate, then, may inadvertently be putting some students at a real disadvantage—a disadvantage that is not necessarily dictated by their ability, but rather by their race, ethnicity, or sex.

To be fair, high-stakes tests aren't just used to gauge school performance. People encounter important tests in many settings—in sports, music, and the business world to name a few. On the one hand, these one-shot snippets of performance are a fast, easy, and efficient way to assess ability. Yet if we end up relying too heavily on tests as the gateway for jobs, team memberships, school admissions, and the like, we may inadvertently limit the talent pool from which we can draw. In many situations, one-shot performances are taken as *only one* metric of an individual's potential. This is in contrast to the high school exit exam—a be-all, end-all hurdle that must be cleared to move on.

At the NFL Scouting Combines, the annual job fair, if you will, for prospective NFL players, athletes are put through a series of drills, tests, and interviews with NFL head coaches, general managers, and scouts each year before the draft. You might think that every test these players take at the minicamp would be considered a very important metric of their skill and ability, but this isn't the case. When prospective draftees get to the camps, interested teams have already compiled a massive amount of data on the players and, for many coaches, the tests the players complete at the camp makes up only one piece of the puzzle. No one test is seen as the holy grail.

The Wonderlic Personnel Tests (often termed the Wonderlic Intelligence Test), for example, has been used in the NFL since 1968 to assess players' ability to think quickly on their feet and make effective decisions under pressure.[19] The test consists of analytical and numerical questions taken in a fast and furious twelve-minute time period.

Players answer questions such as:[20]

1. Look at the row of numbers below. What number should come next?

   8 4 2 1 ½ ¼ ?

2. Assume the first two statements are true. Is the final one:

   The boy plays baseball. All baseball players wear hats. The boy wears a hat.

   1. true, 2. false, 3. not certain?

3. Paper sells for 21 cents per pad. What will four pads cost?

4. How many of the five pairs of items listed below are exact duplicates?

Nieman, K. M.	Neiman, K. M.
Thomas, G. K.	Thomas, C. K.
Hoff, J. P.	Hoff, J. P.
Pino, L. R.	Pina, L. R.
Warner, T. S.	Wanner, T. S.

5. RESENT | RESERVE

   Do these words

   1. have similar meanings,
   2. have contradictory meanings,
   3. mean neither the same nor opposite?

6. One of the numbered figures in the following drawing is most different from the others. What is the number in that figure?

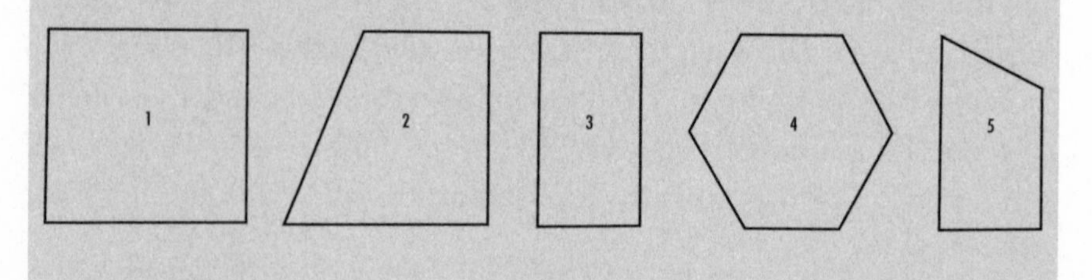

7. A train travels 20 feet in $1/5$ second. At this same speed, how many feet will it travel in three seconds?

8. When rope is selling at $.10 a foot, how many feet can you buy for sixty cents?

9. The ninth month of the year is

   1. October, 2. January, 3. June, 4. September, 5. May.

10. Which number in the following group of numbers represents the smallest amount?

    7 .8 31 .33 2

11. In printing an article of 48,000 words, a printer decides to use two sizes of type. Using the larger type, a printed page contains 1,800 words. Using smaller type, a page contains 2,400 words. The article is allotted 21 full pages in a magazine. How many pages must be in smaller type?

12. The hours of daylight and darkness in SEPTEMBER are nearest equal to the hours of daylight and darkness in:

    1. June 2. March 3. May 4. November

13. Three individuals form a partnership and agree to divide the profits equally. X invests $9,000, Y invests $7,000, Z invests $4,000. If the profits are $4,800, how much less does X receive than if the profits were divided in proportion to the amount invested?

14. Assume the first two statements are true. Is the final one:

    Tom greeted Beth. Beth greeted Dawn. Tom did not greet Dawn.

    1. true 2. false 3. not certain?

15. A boy is 17 years old and his sister is twice as old. When the boy is 23 years old, what will be the age of his sister?

The answers are: (1) 1/8 (2) True (3) 84 cents (4) 1 (5) 3 (6) 4 (7) 300 feet (8) 6 feet (9) September (10) .33 (11) 17 (12) March (13) $560 (14) not certain (15) 40 years old.

Most coaches say that the Wonderlic is only one measure among a litany of physical and mental tests and past-performance results that they use to decide where a player should fall on draft day. The state of California might take a cue from the NFL and do the same when evaluating who should receive a diploma. Rather than just one test, they should allow several different performance measures to dictate whether their students are able to graduate.

One reason that NFL coaches don't place too much weight on the Wonderlic is that scores can readily change with practice. In 1999, when asked about the Wonderlic, San Francisco 49ers manager Bill Walsh told the *Chicago Tribune*, "I think it's pretty obvious it is being manipulated now, so I don't put much stock in it. . . ."[21]

Former Cincinnati quarterback Akili Smith is a good example of the impact that practice can have on a player's Wonderlic score. Smith was the number-three NFL draft pick in 1999. He scored a 12 on the Wonderlic the first time he took it (well below the test average of 21) and an impressive 37 the second time around. This score change created quite a stir in the sports community. If the Wonderlic is supposed to be a stable measure of intelligence, how could a player's score change so quickly as a result of a second testing? Scores from this one test may not tell the entire story of a player's mental ability.[22]

NFL coaches seem to have done a good job of putting scores on the Wonderlic in perspective. Other institutions should do so with scores from tests they use. In other words, we know practice on these tests makes a big difference in students' scores, so it may not be that these tests are truly assessing all students' abilities equally.

Like the Wonderlic, significant practice effects on tests like the SAT or GRE can occur. In fact, there is a huge market for test-preparation courses where practice taking actual tests is one of the big selling points. Test-prep classes like Princeton Review or Kaplan not only teach students tricks to get through particular test problems; these classes also give students lots of practice with real, timed tests. Practicing under the types of pressures you will face on the big testing day is one of the best ways to combat choking, but not everyone has the opportunity to take test-prep classes. These classes are very expensive.

A six-week SAT test-prep class in Chicago, for example, will run a student about a thousand dollars. High-stakes test scores may not be telling us everything we want to know, because some students miss out on the opportunity to get ready for the test, which limits their ability to perform at their best when it counts the most.

> Practicing under the types of pressures you will face on the big testing day is one of the best ways to combat choking.

Minorities are not even taking some tests, such as Advanced Placement or AP exams, which help high school students earn college credit before they have matriculated to the university level. Although the program has grown steadily over the last five years—with almost 800,000 students from U.S. public high schools taking at least one AP exam in 2009—the gap in who takes the tests is still wide.[23] Here are the numbers: 14.5 percent of high school graduates in 2009 were African-American, but only 8.2 percent of students taking the AP exams were black. In contrast, white high school students represented about 61 percent of all 2009 graduating seniors and took the AP tests at roughly the same rate, which means that white students are taking the AP tests at a rate proportionate to their representation in high school.

Many people argue that high-stakes performances are precisely the snippets of ability that should be used to identify talent. After all, don't we want to find out who can perform best under stress? But if high-pressure situations don't affect everyone equally, and the opportunities to learn to deal with the stressors are not available to all, then those who do have the talent or skill to succeed may never have the opportunity to do so. Some students are not even taking the tests in the first place. Pressure is not an equal opportunity terrorizor. Important test scores and do-or-die performances don't always give us a good idea of who has what it takes to succeed.

So, what do we do about it? One solution is to downplay the obsession our culture has with testing, a step many teachers, psychologists, and administrators have been pushing for some time. Indeed, my research group and I have pointed out that the very pressure-filled situations in which standardized tests are administered are a big part of the problem—because these high-stress situations push the best students downward. Others have echoed these claims, pointing out that the inability of the SAT and GRE to accurately reflect the performance of certain underrepresented groups, such as high-achieving women and minorities, is especially troubling. Indeed, even the National Math Panel, convened by President George W. Bush to help keep America competitive in math, has acknowledged the role of math anxiety and stress in derailing the performance of those who in less pressure-filled situations show potential for success.

Of course, as long as people are so focused on finding one measure, one test, one score to use in making admissions, scholarship, and job decisions, then it is important to focus on what we can do to ensure optimal performance when it is needed most. In chapter 6 we tackle the issue of how to alleviate choking under pressure in the all-important high-stakes situation or test.

CHAPTER SIX

THE CHOKING CURE

Early one January morning a few years back—on one of those horrifically frigid Chicago days when even the dogs don't want to go outside—I received an e-mail from a woman at the California Department of Education inviting me to fly to Southern California the following month to give a talk to a group of math educators. Teachers from grades K–12 and administrators from almost every school district in the state would be gathering over the long Presidents' Day weekend to discuss the status of standardized test performance in the California public schools.

The meeting's organizers had put forth some pretty lofty goals for the three-day conference. By meeting's end, they aimed to come up with a set of concrete tools that would be quick, easy, and cheap to use in the classroom to improve the math test scores of the state's students. Of particular interest were students' scores on the annual achievement tests that were taken at each grade level, because low scores would jeopardize federal education funding to the state. They also wanted to improve scores on college assessment tests such as the SAT and ACT. High scores on these tests would help California students gain admittance to the best colleges and universities in the country, which would, of course, reflect well on the public school education system in the Golden State.

Born and raised on the West Coast, I try to travel back to California often, even though I have made the Midwest my home for the better part of fifteen years. The day-in, day-out cold, gray skies, and the never-ending snow in the dead of winter can be a lot to handle, so whenever I get the opportunity to spend a few days somewhere warm, I try to take it. It also seemed especially important to share what we psychologists know about enhancing test performance with the teachers who spend their days getting kids ready to take high-stakes tests.

Each year, the U.S. Department of Education puts out a national report card that ranks the fifty states according to their students' academic achievement.[1] California is consistently near the bottom of the list. Now, to be fair, the Golden State's numbers don't always hit rock bottom: One number for which California is at the top represents the magnitude of the achievement gap between white and black students. California has one of the biggest racial achievement gaps in all of the United States. From elementary school through high school, white students in California consistently perform at a substantially higher level than their minority counterparts. Like your golf score, this is a number that is better when it's lower. I couldn't say no to an opportunity to help raise California's rankings in achievement while at the same time lowering the racial achievement gap. And, yes, swapping my snow hat for sunscreen seemed like a nice bonus. So I accepted the invitation and got to work gathering the research that I was going to share with the educators who would be attending the Southern California meeting.

I arrived in San Diego the night before I was supposed to speak at the conference and headed to a dinner that had been arranged for the meeting's speakers at the hotel's main restaurant. I sat down next to a late twenty-something guy named Alex, who, I soon learned, was in his second year teaching tenth-grade algebra at Union High School in

San Francisco, and taking the night classes he needed to get his master's degree in education at the same time.

Alex's principal had suggested he attend the California math conference to get to know other teachers in the state and the math education community. Attending the meeting was a welcome reprieve from his hectic daily schedule and being at the meeting would also earn him credit that could be counted toward his education degree. Alex told me how relieved he felt to be close to finishing his dual life of teacher by day and student by night. It wasn't so much the crazy amount of work he had—although it was a lot. He was really looking forward to a reprieve from the stress of his first few years on the job. He was constantly worrying about whether he was cutting it in the classroom and he felt stress had taken a toll on his ability to function at his best. Alex told me he felt like his brain just wasn't all there lately. His impressions that his brainpower was depleted were probably spot-on. Being overloaded and under stress can take a toll on your cognitive abilities, but, as I told Alex, our brains are pretty good at bouncing back in the long run once the stress is gone.

EXAM TIME

A few years ago, a group of psychologists at Cornell University's medical school got hold of two dozen medical students who were spending the better part of a month preparing for their intensive exams.[2] Passing these exams meant advancing in medical school. Failing put students on the road to a quick exit from medical school and everything these students had worked so hard to achieve.

The medical students had been convinced to take a break from their studying and spend a few hours doing attention and memory tasks while their brains were scanned using fMRI. The psychologists were also going to scan another group of people who were the same age as the medical students, had the same sleep habits, and similar years of education. The big difference between them was that this second (control) group of people was not facing the gloom and doom of year-end medical exams.

The cognitive tasks everyone did in the fMRI scanner were fairly simple, but the stressed-out medical students performed poorly on them. The medical students were sluggish when they had to switch from identifying the color of an object presented on the computer screen (say, a red triangle) to identifying which direction it was moving. The medical students were easily distracted from whatever task they were doing in a way that the nonmedical students were not. When the researchers peered inside everyone's brains to see how they were functioning, they found that the stress that the medical students were feeling was reducing the cooperation of different parts of the brain that usually worked together to support thinking and reasoning. In particular, the prefrontal cortex didn't seem to be working as hard for the medical students and was not as in sync with the rest of the brain (for example, parts of the parietal cortex) as it should have been.

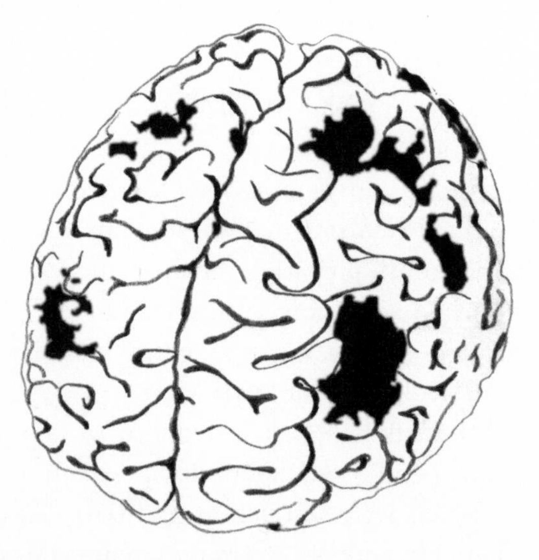

A depiction of the frontal-parietal working-memory and attention brain network that became decoupled under stress. The front of the brain (that sits above the eyes) is located toward the bottom of the picture.[3]

The prefrontal cortex, among its many functions, houses working-memory. People who can use their prefrontal cortexes to the greatest degree are highest in cognitive horsepower. But the medical students, who did have high-functioning working-memory, were not using their powerful brain resources to their full potential, most likely because of the stress they were under. This was probably the same reason why Alex, the teacher, wasn't feeling fully functional—the stress was tamping down his brain function.

Alex looked discouraged by this information, but I quickly explained that the Cornell study also had shown that the effects of stress on the brain are reversible. A month or so after the medical students took their year-end exams, their brains were scanned again. This time the medical students' brain functions looked just like the non-stressed-out control group as they performed the demanding thinking and reasoning tasks.

These results are intriguing because they reinforce our understanding of the ways that stress changes the brain. Being under pressure alters how different areas of the brain communicate. In a nutshell, the prefrontal cortex works less well and decouples—or stops talking to—other brain areas that are also important for maximal cognitive horsepower. The brain generally works in concert, as a network. When a particular brain area stops communicating with other areas, this can have dire consequences for our thinking and memory capabilities. Just think about what happens in the middle of a heated argument with your spouse. All of the sudden your head is not as clear as you want it to be and you just can't seem to make your point. Some of the stress you are experiencing may be preventing your brain from working as a network, and as a result, you have trouble coming up with the best examples to support your viewpoint.

Fortunately, the brain seems able to rebound after the stress is gone, although this doesn't happen right away—once the worries commence, they don't immediately subside even after the stress-inducing situation is over. But after some time in nonstressful environments, our brainpower does seem to recover. Of course, this recovery is only possible if we don't always feel under the gun. That may be one reason we seem to come up with the good one-liners or counterarguments for our spouse long after the stressful argument is over.

Alex was happy to hear that he would return to normal once the pressure was off and once he felt more comfortable in his role as a teacher. But for some people, the stress may not go away so easily. Think, for example, about the stress on a woman who is one of the few female graduate students in the Ph.D. program in mathematics at Stanford University. If she constantly walks around with fears about her ability to succeed in a male-dominated field, her stress may never subside. Even the physical environment will keep her stress high, since every time she has to go to the bathroom, she has to make a trek down to the basement to use the only women's bathroom, a constant reminder that she doesn't fit in. Or what about a Latino accountant at an all-white firm who has to combat stereotypes held by both coworkers and clients about his credentials and the intelligence of his race. He is likely in a permanent, stressed-out state. If a mere month of studying for an exam can have such a big impact on medical students' cognitive horsepower, just imagine what constant stress does to people's brains.

A THREAT ALL AROUND US

From behind the podium at the California meeting I asked the two hundred or so teachers how many had heard of *stereotype threat.* I felt a little silly asking the question because Claude Steele, the professor who had coined the term a few years earlier, was just up Highway 1 at Stanford University.

Only a handful of teachers raised their hands, so I asked, How many of you have seen instances in the classroom where students underperform relative to their potential merely because these students feel discouraged about their ability to succeed? This time almost everyone's hand went up.

I explained that one reason we think this underperformance occurs is that students are aware of stereotypes about the intelligence of their sex, race, and ethnic group. In fact, this poor performance may be a major contributor to the racial achievement gap in California, where minority groups such as African-Americans perform at a lower level, on average, than white students in the classroom and on standardized tests. There are some surprisingly simple things that can be done to shrink this difference, however.

Several years ago, psychologist Geoff Cohen and his research team at the University of Colorado went into seventh-grade classrooms with an intervention designed to reduce the racial achievement gap.[4] Much like a randomized-control drug trial in which doctors and their patients are blind to who is part of the treatment and who is merely getting a placebo pill, neither the teachers nor the students knew much about what was going on. The intervention was simple, fast, and required little money or classroom time and ended up boosting minority student performance more than anyone—even the researchers themselves—ever imagined.

Cohen and his group targeted a suburban public middle school in the Northeast that consisted of mainly middle- and lower-class students. It was the perfect place to conduct the work because the student body was about equally divided between African-American and white kids—and, unfortunately, there was also a strong gap in how the black and white students performed in the classroom.

Early in the fall term of the school year, the researchers asked teachers to administer an "exercise packet" to all of the seventh graders. From the outside all the packets looked exactly the same but there were actually two different sets of packets. Students were assigned, at random, to receive either a treatment or a placebo packet. Importantly, the teachers didn't know which group the students were assigned to and the students didn't even know that the exercise packet they were completing was part of an intervention in the first place.

When the students opened their packets they found a piece of paper listing a bunch of different values that people might hold. For instance, some might value their relationships with their friends or family, being good at art, or their athletic ability. Students in the treatment group were asked to indicate the value that was *most* important to them and to write a brief paragraph explaining why they thought this value was an important one. Students in the placebo group were asked to indicate the value that was *least* important to them and to write a paragraph about why this value might be important to someone else. Once students had finished writing, they were instructed to place their packet back in the envelope it had come in, seal it, and return it to their teacher. The teacher then resumed his or her lesson plan. The entire procedure took about fifteen minutes total.

At the end of the fall term, the researchers were given access to the official transcripts of all of the students. Black students overall still performed worse than their white counterparts, but the black students in the treatment group (the students who wrote about their most important values) performed better than the black students in the placebo group by about a fourth of a grade point. This improvement was not limited to a few students—the exercise packet benefited about 70 percent of the black students in the treatment group. Among the white students, meanwhile, there was no difference in performance between the treatment and placebo groups.

Given that the average difference in academic performance between black and white students was about 70 percent of one grade point, and that the blacks in the treatment group improved by roughly 25 percent of a grade point, the results of this study represent almost a 40 percent reduction in the racial achievement gap.

Was this result a fluke? To answer this question, Cohen and his colleagues ran the same study a year later with different students. They got the same results. The likelihood of observing their treatment effects—twice—by chance is about 1 in 5,000. So why did this simple fifteen-minute exercise have such a big effect? The answer seems to lie in the phenomenon of stereotype threat. Most of the teachers I was speaking to had seen it affect their students but didn't know that it was being systematically studied by psychologists around the country.

In chapter 5, we saw that when African-Americans are pushed to think about negative stereotypes about their intelligence, they perform poorly on achievement tests. It's likely, then, that minority students who know about negative stereotypes are affected by it. This knowledge prevents students from devoting their full attention and effort to what they are doing, so their grades and test scores suffer.

Cohen thinks that one way to reverse this weight of negative intelligence stereotypes is to allow black students to reaffirm their integrity. Having African-American students write about qualities that are important to them may enhance their sense of self-worth and value, which may in turn buffer them against negative expectations and their consequences. Although this writing assignment only happens once, it can have a big effect—students do better academically after they write, which reduces the power of the stereotype. A negative cycle is changed into a positive, reinforcing cycle.

These effects seem to last. Cohen and his colleagues completed a two-year follow-up study with their first batch of students and the treatment that the black students had received in seventh grade continued to exert a positive effect on their performance throughout middle school.[5] The likelihood that a black student would enter one of the school's remedial programs or repeat a grade during the two years after the intervention was much smaller if the student had been in Cohen's treatment group and had written about self-affirming qualities, compared to the placebo group where the students wrote about values important to other people. Likewise, the percentage of black students who were put in Advanced Placement courses after Cohen's intervention was higher for those students who performed the treatment writing as opposed to those in the placebo group.

Most of the teachers at the Algebra Forum seemed impressed with these findings. The raised performance bears an implication that is difficult to refute: the racial achievement gap is not an intractable phenomenon—and it does not take millions of dollars or hundreds of worker hours to begin to change it. You can start by simply setting aside a few minutes for students to reflect on their positive qualities.

Of course, it's not *always* the minority students who need help in school or on important tests. You might wonder if there is anything that will help to boost test scores for all students—regardless of ethnicity, sex, or race. My research team and I have wondered the same thing and have done work that suggests the answer is yes.

WRITING CAN DO WONDERS

The intervention performed by Geoff Cohen and his colleagues suggests an easy way to help boost minority student performance in the classroom. If you constantly walk around with a cloud of self-doubt over your head—brought on by the mere awareness of stereotypes about your intelligence—then writing about your self-worth might very well shrink that cloud. As a result, school performance improves, the self-doubt shrinks again, and so on. But what if a student isn't worried about racial stereotypes per se, but still carries a weight on his or her shoulders? This weight might be the result of the expectations and pressures most students face in today's competitive academic environment. Can we do anything about that?

You have no doubt heard that getting a problem "off your chest" will make you feel better. You may even have experienced the cathartic effect of talking about your worries. New research out of my Human Performance Laboratory shows that disclosing does more than just make you feel better: it may actually change the workings of your brain when the pressure is on. The end result is better test performance under stress.

My graduate student Gerardo had been closely following Cohen's work with the seventh graders for some time and figured that one reason the intervention worked so well was that writing about one's self-worth helped to deflect worries about the negative and promote engagement with school. If so, then Gerardo thought it was a small stretch to imagine that a general writing exercise that helped people to deal with their worries might serve as a performance boost for all types of people—extending beyond blacks' own performance in the classroom.

Gerardo asked a high-achieving group of mostly white University of Chicago students to take a difficult math test while we ratcheted up the stress. We called in our usual techniques for putting the pressure on—including offering students twenty dollars for stellar performance and reminding them that, if they performed poorly, they would jeopardize the ability of a partner who also wanted to win money. We also videotaped students and told them that math teachers and professors would be watching the tapes to see how they performed. [6]

Immediately after telling them what was on the line, we asked some students to write for ten minutes about their thoughts and feelings concerning the test they were about to take. We wanted the students to get their feelings about the pressure off their chest so we told them that they couldn't be linked to their writing by name so they should feel free to write openly and freely about any worries. Other students were not given the opportunity to write, but just sat patiently for about ten minutes while the experimenter got all the testing materials together. What we found was quite amazing. Those students who wrote for ten minutes about their worries before the math test performed roughly 15 percent better than the students who sat and did nothing before the exam.[7]

Keep in mind that this difference doesn't just reflect variation in math ability across our writing and no-writing groups. We know this because everyone took a practice math test before the experiment got started and there was no difference in performance between the two writing groups. Those students given the opportunity to write about their worries before the pressure-filled math test aced it, while those who didn't write choked under the pressure. Writing about your worries before a test or presentation prevents choking.

How could writing down your worries stop your choking? You might have expected the exact opposite to occur given that people were reflecting consciously on the upcoming pressures and that increasing the salience of the

> Writing about your worries before a test or presentation prevents choking.

pressure could lead to poor test performance rather than better performance. But we are not alone in finding a benefit in writing about anxieties and worries.

For several decades, psychologist James Pennebaker has been extolling the virtues of writing about personally traumatic events in your life, such as the death of a close family member, a difficult breakup, or leaving home for the first time to attend college. One reason Pennebaker is such a fan of writing is that he and his colleagues have found, time and time again, that after several weeks of writing about a life stressor, people have fewer illness-related symptoms and even show a reduction in doctor's visits.[8]

Expressing your thoughts and feelings about an upsetting event—whether a trauma in the past or a pressure-filled test coming up in the future—is similar to "flooding therapy," which is often used to treat phobias and post-traumatic stress disorder. When a person repeatedly confronts, describes, and relives thoughts and feelings about his or her negative experiences, the very act of disclosure lessens these thoughts. This is good for the body because the chronic stress that often accompanies worrying is a catalyst for health problems.

Although the benefits of flooding therapy are sometimes questioned because psychologists worry that the flooding actually anchors the body and mind to the trauma and reinforces it so it's harder to get over it, the gold standard for dealing with traumatic events in the long term does involve repeatedly and intentionally bringing the traumatic event to mind. Trying to forget, block, or suppress the emotional incidents without attempting to reorient or reappraise the distressing event in the first place has not been proven in psychopathology research and practice to be an effective way to deal with emotionally charged memories.[9]

Simply put, disclosure seems to be good for the body and for the mind. When university freshmen, for example, are asked to write about the stress of leaving home for the first time and going off to college, they report a decrease in their worries and intrusive thoughts. Writing about their worries also leads to improved working-memory over the course of the school year.[10] Expressive writing reduces negative thinking, which frees up cognitive horsepower to tackle what comes your way.

The written word can be so powerful because, according to psychologist Matthew Lieberman at the University of California, Los Angeles, putting your feelings into words changes how the brain deals with stressful information. Lieberman discovered the power of words a few years ago when he and his research team asked people to look at pictures of emotional faces and pick the word—such as *scared* or *angry*—that best described the face's expression.[11]

Putting your feelings into words changes how the brain deals with stressful information.

When people simply looked at the emotional faces, they showed a lot of brain activity in areas such as the amygdala, which is involved in our emotional experiences and reactions. When the amygdala is highly active, it can prevent other areas of the brain needed to bolster cognitive horsepower from working their hardest. However, when people viewed the faces and picked the word that best described the face they saw—in other words, when they labeled the emotional face with a word—activity in the amygdala was dampened. Indeed, using words to label the faces led to increased activity in brain areas such as the prefrontal cortex, which in turn seemed to reduce the response of the amygdala, thus helping to alleviate people's emotional distress.

The positive brain changes that were evident from labeling the emotional faces with a word were not apparent when people viewed emotional faces and were asked to choose another face displaying the same emotion, without articulating a word that best described the emotion. It's not just understanding emotion that helps to thwart the overreaction of emotional centers of the brain; it's actually putting these feelings into words that does the trick.

Disclosing negative information and labeling it as such frees your mind from unwanted thoughts and helps you focus on something other than the negative. In addition to written expression that can help people thrive under pressure, psychologists and neuroscientists have recently discovered that the age-old practice of meditation helps extinguish distracting and worrisome thoughts that plague us in stressful times. Experienced meditators can clear their minds of unwanted information when practicing their ancient art, but new research suggests that, even when not meditating, people who practice this tradition are better than those who don't at facilitating their brains' return to a calm, cool, and collected state when stressful events come their way.

THE CALM, COOL, AND COLLECTED BRAIN

University of Wisconsin neuroscientist Richard Davidson has practiced meditation since the 1970s, but only recently started to study it. Advances in brain-imaging techniques (for example, fMRI) have allowed him to take a look inside the brains of meditators and nonmeditators to see how they differ. Davidson had a hunch that meditation has a powerful effect on the brain and now research has confirmed it.[12]

Recently, a dozen people who had been practicing Zen meditation daily for several years and people who didn't meditate partook in a watered-down form of meditation while their brains were scanned using fMRI.[13] Often in Zen meditation, people focus their full attention on one idea or object (such as their breath) and clear their mind of everything else. While lying in the scanner people looked at a blank computer screen in front of them and were told to focus on their breathing. Every once in a while, a word appeared on the screen (say, *jacket*) and the people were to decide whether it was indeed an English word, after which they were to promptly "let go" of what they had just read and refocus again on their breath. The sudden appearance of the words was designed to mimic the appearance of spontaneous thoughts.

Researchers conducting the Zen meditation study wanted to

know if meditation practice changed people's ability to recover from the intrusion of a word that unexpectedly appeared on the computer screen, which would strongly suggest that meditation improved recovery from the distractions induced by spontaneous thought as well. After the interruptions, the experienced meditators' brains did indeed return faster to a relaxed state than nonmeditators' brains. Intensive meditation practice seems to reduce the elaborative thinking that normally occurs when we evaluate a thought and meditators are able to clear their minds of distractions more quickly than those who don't meditate.

Obviously, this type of thought control could be useful in dealing with worries that arise in stressful testing situations. Worries about screwing up often cause flubbed performances because they use up valuable working-memory resources that could otherwise be devoted to test taking. By training the brain to discard negative thoughts, you can thwart the negative effects of stress. Discarding these thoughts is not the same as trying to ignore them or dismiss or suppress them, which uses up working-memory. When a worrisome thought arises, you acknowledge it, name it (as if actually identifying it and writing it down), but then let go of it. You don't attach any more brainpower to it.

You don't even have to practice meditation for several years to reap the thought control benefits that it brings. A recent study by Davidson and his colleagues showed that only three months of intensive Vipassana meditation (a practice in which you observe whatever thoughts and perceptions arise in your consciousness without making negative judgments about them) reduced people's tendency to have their attention captured by unwanted thoughts or events.[14] Davidson's research was on a phenomenon we psychologists call the "attentional blink." When two things are presented in close succession, people often get hooked on the first item and can't pull themselves away to attend to the second item. You can see how a student could get stuck on a negative thought under pressure and "blink" right past the test problem in front of him, or at least give the problem less attention than it deserves.

Davidson showed that a three-month course in meditation helped people limit their focus on the first of two items presented to them so that they were more likely to notice the second one. The people in Davidson's study didn't have years of meditation experience, so even a small amount of meditation training helped reduce their propensity to "get stuck" in distracting thoughts or activities. Essentially, meditation training allows people to develop means to engage and disengage from what they experience—something that is extremely useful for battling self-doubt in pressure-filled situations.

Recently, in my laboratory, we have shown that people with no meditation experience at all can benefit from about ten minutes of meditation training before they take a pressure-filled math test.[15] College students who were given a short tutorial in mindfulness before they took a high-stakes test scored on average a B+ on the test (87 percent) while those students that didn't get the mindfulness training beforehand scored a B− (82 percent). This difference, although small, is remarkable given that both groups performed similarly on a practice test taken before the stress ensued.

Tiger Woods is an avid proponent of Buddhist practices, including meditation. Maybe this is why he appears to swing the golf club so effortlessly with distractions from the fans and press all around him. Woods has learned a nonreactive awareness of these experiences, and by practicing letting thoughts and feelings come and go as they please, he has likely produced long-lasting changes in his brain function. Of course, even as one of the greatest golfers of all times, you are going to have your ups and downs. Tiger Woods can have an off tournament or perform poorly when the pressure is on just like anyone else on the PGA tour. And even Buddhist meditation practice won't make you invincible to screw-ups in all facets of life.

Phil Jackson, coach of the Los Angeles Lakers, became well-known when he was coaching Michael Jordan and the Chicago Bulls to several successive championships, for advocating meditation practices as a means of enhancing his players' performance. Successful individuals ranging from board members of Goldman Sachs to the chairman of

Ford Motor Company, William Ford, have also touted the benefits of meditation practices in their personal and work lives.[16] These powerful figures in sports and business are often using their intuition about the psychological benefits of their practice, but as we have seen thus far, brain research suggests that these intuitions are spot-on. Holding on to thoughts and worries under stress leads to an inability to perform the tasks you are faced with, and learning how to control your mind so that you are able to hone your attention to what matters (and only what matters) is a real key to success under pressure.

Indeed, actresses from Heather Graham to Goldie Hawn are avid meditators. Hawn even has a dedicated meditation room in her house filled with calming things designed to help her sit quietly, but more important, quiet her mind. Al Gore and Hillary Clinton, politicians who constantly perform under pressure, have attested to the power of meditation in helping to reprogram the mind. The idea that the brain and body can change as a result of meditation is intriguing because it means that meditation can help you perform at your best especially when you need it the most—under stress.[17]

THINKING DIFFERENTLY

I was just about to introduce another technique for promoting academic success under pressure to the teachers at the California math conference when I noticed that my own parents had walked in and were now sitting in the back of the room. Yes, that's right, my parents. When they had learned that I was going to be in California for a talk, they immediately arranged to fly down from San Francisco to hear me speak. Like any good baby-boomer couple, my parents never missed an opportunity to watch me and my siblings on the soccer field or in school plays as we were growing up, and their support hadn't subsided now that we are all gainfully employed adults. My parents' presence was usually a given whenever I was back on the West Coast giving a talk, but in the last several years they had also traveled as far as Europe and Australia to attend my seminars.

I love their support, but it can also be a bit distracting to have them in the audience. When you see your dad smiling at you from the back of the room, it can be hard to remember that you are not just someone's kid, but an accomplished academic as well. If and when this does happen, I immediately think about my research credentials, a trick I developed after discovering that getting people to think about aspects of themselves that are conducive to success can actually be enough to propel them to a top performance and prevent choking.

Getting people to think about aspects of themselves that are conducive to success can be enough to propel them to a top performance and prevent choking.

For example, when Asian female college students are about to take a math test but are first asked to fill out a survey that highlights their Asian identity ("How many generations of your family had lived in America?"), they score higher on the test than if they were instead asked questions that prompt them to reflect on the fact that they are female ("Do you prefer co-ed or same-sex residence halls?").[18] Just thinking about an aspect of yourself that is associated with better versus worse performance can be enough to change how you perform. And when a female college student is asked to describe several different facets of herself—to give a complete description of herself as a woman, athlete, friend, family member, artist, and actor—she is less likely to screw up on a high-stakes math test than if she weren't asked to think about all her complexities.[19] Seeing yourself from multiple perspectives, not just as a girl in a situation where women don't usually excel, can help thwart the spiral of self-doubt and worries that interfere with people's ability to perform at their best.

Mapping out your multifaceted nature can work across the board, for anyone. Just think about the good it might do if a high school student were to spend five minutes, before taking the SAT, drawing a diagram of everything that makes him who he is. Maybe he tutors underprivileged children, manages his basketball team, and can eat more hot dogs than any of his friends in a single sitting. The mere act of realizing you aren't just defined by one dimension—your SAT score or a speech or a solo—can help curtail worries and negative thoughts. In essence, thinking about yourself from multiple perspectives can help relieve some pressure that you feel.

Let's take a step back for a moment and reflect on what happens when students arrive to take a standardized test such as the SAT. One of the first things that test takers do is to check off boxes to indicate their race, their sex, GPA in school, and even their families' income levels. Providing this information can undermine the students' self-confidence, especially if they feel pigeonholed into a group that is stereotyped as academically challenged or unsuccessful. The consequences of filling out this information for test performance can be dire.

Indeed, psychologists Kelly Danaher and Christian Crandall at the University of Kansas found that simply moving the standard background questions about sexual identity from the beginning to the end of the test led to significantly higher performance by women on the AP calculus test.[20] Extrapolating from these AP calculus test findings alone, the researchers estimate that, each year, an additional 4,700 female students would receive AP credit that could advance their standing in college math classes if questions about test takers' sex always came at the end. If the Educational Testing Service (ETS)—the company that makes its livelihood producing standardized tests including the SAT, GRE, and the Advanced Placement (AP) tests—were to implement this simple change in all its tests, I believe that scores would markedly increase across the board.

The teachers in the audience looked impressed. In my experience, educators as a group are not particularly fond of standardized tests—especially since, for many, their jobs depend on how well their students perform on these tests. When the stakes are so high, teachers end up teaching to the test, students learn less overall, and that one score and test becomes a looming measure of everyone's success. If teachers didn't feel that their classroom's performance was a measure of their own teaching ability or if students didn't feel as if their test scores dictated how smart they were, students might actually test better. Getting a student to consider himself or herself from multiple perspectives can help, and moving the questions that ask for information about sex, race, or family income to the back of the test works well, too. It's all about taking the emphasis off of the test and letting go of the idea that one particular score or grade reflects a student's intelligence, self-worth, or even his or her potential for greatness.

SEEING IS BELIEVING

Over a three-month period during the run-up to the 2008 presidential election, psychologists at several universities across the country asked blacks and whites to take standard GRE tests.[21] In sum, almost five hundred Americans completed the exams. The test takers were not trying to gain admittance to a graduate program in psychology and actually knew very little about why they were taking the tests. The test takers, out of the goodness of their own hearts, had volunteered for the study because the researchers had asked for their help.

On an initial test that was given prior to the Democratic National Convention (DNC), and so when Barack Obama had not yet accepted his party's nomination for president, white test takers scored better on average than their black counterparts despite the fact that the education levels of the whites and blacks in the study were equated. This finding represents the racial achievement gap that can be found starting in elementary school and that continues all the way through the university level in the United States. However, on a test administered immediately after Obama's nomination acceptance speech, and on a test given just after Obama's presidential election victory, things changed. Black test takers' performance improved to such an extent that their scores no longer differed substantially from whites.

As Obama took center stage as the next leader of the free world, the worries that blacks felt about their ability to perform well on a test designed to gauge intelligence seemed to have disappeared. Certainly the stereotype that blacks are not as intelligent as whites can't be true if a black man can become president of the United States. Seeing that these stereotypes were untrue, students didn't worry about confirming them in the first place.

> On a test administered immediately after Barack Obama's nomination acceptance speech, and on a test given just after Obama's presidential election victory, black test takers' performance improved to such an extent that their scores no longer differed substantially from whites.

Only time will tell how wide-reaching this Obama effect is. One thing is for sure: this type of "modeling" isn't just limited to the president. A growing body of evidence demonstrates that seeing can be believing for people who are worried about their abilities because they belong to a racial, sex, or ethnic group that has been negatively stereotyped.

A few years ago, psychologist Nilanjana Dasgupta interviewed women at the start of their freshman year at two liberal arts colleges in the same town in the eastern United States. Both colleges attracted students with similar backgrounds, but there was one big difference between the two schools: one of the colleges that Dasgupta worked with was coeducational and one was not.[22]

Dasgupta interviewed around eighty female first-year students at the two colleges about their views regarding women's abilities to succeed in leadership roles. In short, students at both schools believed that the qualities that women often possess, such as compassion and caring, generally made them better suited to be followers than leaders, a view that is endorsed by most Western societies. However, when the same women were interviewed about the same topic, as sophomores only twelve months later, the female students at the all-women's college had completely abandoned their beliefs that men were better suited for leadership positions. But those women who were now sophomores at the coeducational college actually expressed even stronger views about men being a better fit for positions of power than they had the year before. The women from both schools had possessed similar beliefs about the traditionally different roles for men and women when they entered college, but their beliefs had diverged substantially in different directions one year later. Why?

Surprisingly, the answer has nothing to do with being at a coeducational versus all-women's college per se. Rather, it has to do with differences in the numbers of women in leadership, executive, or managerial positions to whom students are exposed at these two schools. In general, female students at all-women's colleges see a lot more women in leadership roles—from college deans to female professors—than do students at coeducational institutions. And students' exposure to women in leadership roles changes their views about women's suitability for positions of power. We know this because Dasgupta was able to show that the change in students' attitudes about their own sex and their potential for leadership over the course of their first year of college—regardless of what school they attended—was completely driven by the number of female faculty the students had. When female students are exposed to female faculty their attitudes about women change. And the likelihood of students having female faculty is higher at an all-women's school.

Even more amazing is that these exposure effects don't just extend to students' attitudes. They can also affect how well female college students do in math and science classes. For female students at the U.S. Air Force Academy, for instance, having female professors in introductory math and science courses leads to higher grades—especially among those students scoring toward the top of the class to begin with. And if a female student's introductory math and science professors are exclusively female, she is much more likely to major in a science field than are female students with exclusively male faculty in these same courses.[23]

When students attend the Air Force Academy, they cannot choose their introductory math and science courses or the professors teaching them. All faculty teaching the same class use an identical syllabus and give the exact same exams during a common testing period. Because of this, students cannot self-select into (or out of) courses or choose certain professors over others. Whenever a female student is lucky enough to be paired with a female professor in her math and science courses, however, she performs substantively better in these disciplines than female students who are not so paired.

One reason this may be the case is that when women are reminded of stereotypes about their ability (it seems likely that stereotypes are always in the air at a place like the Air Force Academy), *but* are also shown that the stereotypes are not necessarily indicative of women's true skill (either by explicitly being told the stereotypes don't apply or by seeing women who have been able to overcome these stereotypes), poor performance under stress is curtailed. When female students took a math test as part of a study purportedly looking into differences between the sexes in math, but were first told that "it's important to keep in mind that, if you are feeling anxious while taking this test, this anxiety could be the result of these negative stereotypes that are widely known in society and have nothing to do with your actual ability to do well on the test," their performance did not differ from men's.[24] But when women are not given information that undermines the credibility of negative gender stereotypes in math (either because they don't see women who defy these stereotypes or because they are not explicitly educated that the stereotypes are not well founded), women's math performance suffers relative to men's.

Unfortunately, given that only a very small percentage of math and science full professors in the United States are women, most female students aren't frequently exposed to women faculty in these disciplines. As a result, some women's groups have taken it upon themselves to help female students meet women in prominent positions in academia and business. Women receive over half the undergraduate college degrees in the United States, yet this percentage dwindles at the master's and Ph.D. level and continues to decrease as the prominence of positions—especially in math and science—increase.[25] The hope is that female students who meet other women who have defied the odds may help to change these numbers.

After Larry Summers's 2005 remarks, professors at Rice University in Houston began organizing an annual conference where promising female graduate students and postdoctoral fellows came together to learn from successful women faculty about the skills needed to excel on the job hunt, as an assistant professor, and all the way through their academic careers.[26] The female students attending the workshop also learned about the myriad issues they may face as they climb up the academic ladder, which include everything from unconscious sexism to the two-body problem (finding a job for your spouse, especially if he or she also has scientific ambitions), but the most important part of the program likely comes from being able to interact with exemplars of female academic success.

These programs are popping up as early as elementary school. Take the Winchester Thurston School in Pittsburgh, Pennsylvania.[27] Kelly Vignale, a WT science and technology teacher, founded a program called "L3" or "Ladies Who Lunch and Learn" for the WT elementary school. Each week, a different female scientist comes to have lunch with the elementary school girls to talk about their career. The goal is for these girls to see women in top academic positions in the sciences and to get them interested in the work going on. L3 shows girls that being a scientist, engineer, or mathematician is not just for boys.

There are programs at the high school level as well, such as the Women in Science and Engineering (WISE) Program, which started in 2005 as a partnership between Garrison Forest School, an all-girls K–12 day school in Maryland, and the Johns Hopkins University. Garrison Forest recruits its own students as well as female students from around the country to spend a semester of their junior or senior year in residence at Garrison Forest. WISE students have a special curriculum at GF where they meet and work with first-rate math and science teachers on their own campus. These students also spend several days a week on the Johns Hopkins campus working in a laboratory doing hands-on research in chemistry, cognitive science, and mechanical engineering, to name a few areas. Not only do these high school women get to conduct research in the flesh, but they are also exposed to the various careers that are available to them in math, science, and engineering fields and to the breadth of role models in these sciences as well.[28]

Another successful modeling program is one started by Marnie Halpern and Lissa Rotundo.[29] Halpern is a biology professor at Johns Hopkins and Rotundo teaches biology at a Baltimore high school. For several years now they have run a weekly Women Serious About Science (WSAS) program, in which female scientists are brought into local high schools to meet with girls to discuss their education and their research interests and to share personal stories of how they got to where they are. At first the organizers called on local friends and colleagues to be the speakers, but since the WSAS's inception, high school girls have heard from engineers, astronomers, chemists, physicists, cancer researchers, and neuroscientists.

The model is simple: Marnie finds the speakers, and a teacher at the high school at which the speaker is going to speak makes the arrangements. Everything is kept pretty informal with the scientist speaking to girls about their experiences over a pizza lunch. Not only do these high school girls get to see women scientists in the flesh, but they also learn about the different types of math, science, and engineering career options that are open to them. The scientists benefit as well because the WSAS program has come to serve as a powerful recruiting tool for researchers looking for talented high school girls to spend summers interning in their laboratories.

WHERE ARE WE NOW?

Testing is a large component of our educational system, and today it's not uncommon for students to take tests even to gain admittance to kindergarten. These tests continue all the way through graduate school and testing to get and keep your job is now becoming more common. If you are able to perform at your best in important tests, future opportunities abound. If not, low test scores can result in poor evaluations from mentors and teachers, lost scholarships, and other educational and employment opportunities.

Performing poorly in pressure-filled testing situations can also lower people's confidence in their ability to succeed and their willingness to venture further in particular areas. A girl who bombs a math test may decide that girls really can't do math and thus choose a path through her schooling that limits her exposure to this subject. It's easy to see how a recursive cycle could emerge here. Poor math test performance leads to avoidance of math classes, which in turn means knowing less math, poorer test performance, and so on.

Here is a summary of quick tools we have covered thus far that can help you perform at your best, as well as a few extra pointers that may help you perform up to your potential in stressful academic situations:

Tips to Ensure Success Under Stress

Reaffirm your self-worth. Before a big test or presentation spend a couple of minutes writing about your many interests and activities. This writing can promote feelings of self-worth. Reaffirming yourself, especially when you question your abilities, can boost your confidence and performance.

Map out your complexities. Before taking an important test, spend five minutes drawing a diagram of everything that makes you a multifaceted individual. This exercise can help to highlight that this one test score doesn't define you, which can in turn take some of the pressure off.

Write about your worries. Writing for ten minutes about your worries regarding a presentation or test you are about to take can thwart the anxieties and self-doubt that often emerge in high-pressure situations.

Meditate away the worries. You can train your brain not to dwell on negative thoughts and instead recognize and then discard them. Meditation training can help you harness all of your cognitive horsepower for the task at hand.

Think differently. Think about yourself in ways that highlight your propensity for success. Instead of thinking, for example, that you belong to a sex or racial group that is unfairly stereotyped to be bad at math, remind yourself instead that you have the tools to excel—maybe you are a college student at a prestigious university or you have done well in school in the past. Focus on your credentials to help turn a bad performance into a good one.

Reinterpret your reactions. If you get sweaty palms and your heart races under pressure, remember that these physiological reactions also occur under more pleasant circumstances, such as when you have met the love of your life. When under pressure, if you can learn to interpret your bodily reactions in a positive way ("I am amped up for the test") rather than negative ("I am freaking out"), you may be able to turn your body to your advantage.

Pause your choke. Walking away for a few minutes from a challenging problem that demands working-memory can help you find the most appropriate solution. This "incubation" period helps you to let go of your focus on irrelevant problem details and instead think in a new way or from an alternative perspective—and can produce an "aha" moment that can ultimately lead to a breakthrough and success.

Educate the worries. Merely drawing attention to the stereotypes students may hold—for instance, "Girls can't do math" or "Whites are not as good at math as Asians"—and reminding them that they are stereotypes and nothing more can help to prevent people from worrying about their ability when the pressure is on. It might seem counterintuitive that teaching people about a stereotype regarding their ability would quell its effects rather than exacerbate them, but giving people an excuse for their worries allows them to see their performance as less diagnostic of their intellect.

The Obama effect. Seeing examples of people who defy common stereotypes about sex, race, and ability can help to boost the performance of people in these social groups. After all, if a black man can become leader of the free world, certainly the stereotype that African-Americans are not intelligent just can't be true.

Practice under pressure. The old adage that practice makes perfect can do with a bit of adjustment. Studying under the same conditions you will be tested under—for instance, in a timed situation with no study aids—helps you get used to what you will experience on test day. There is also research suggesting that testing yourself on material (rather than simply studying it) helps you remember it better in the long term. After all, you are going to be tested during the test so you might as well practice being tested.

Outsource your cognitive load. Write down the intermediate steps of a problem rather than trying to hold everything in your head. This provides you with an external memory source, one that may be relatively free of worries compared to your own prefrontal cortex. As a result, you may be less likely to mix up information or forget important details of what you are doing.

Organize what you know. Take a clue from expert waiters like JC, whom we talked about in chapter 2. Coming up with meaningful ways to organize the information you need to remember for a big test or presentation can help take the burden off your working-memory and actually help you remember more.

A LOOK AHEAD

In the next several chapters we move away from the academic world and look at choking in the sporting and business worlds. Stressing out when you have a pencil in your hand does have some similarities to the pressure you may feel when you are wielding a golf club, but there are important differences, and these differences carry implications for how to medicate a choke.

We will also look into the universality of the clutch player. Do the same people shine in the boardroom and on the playing field in high-pressure situations? Finally, we will use what we know about why people choke under pressure to identify tools to prevent or alleviate botched performances in sports and other activities more generally.

CHAPTER SEVEN

CHOKING UNDER PRESSURE

FROM THE GREEN TO THE STAGE

It's pretty easy to find top-ten lists of the greatest "chokes" in sport history when trolling the Internet—or to compile your own. The 2004 New York Yankees usually make the cut. In the fall of 2004, the Yankees were up three games to zero against the Boston Red Sox in the American League Championship Series. Everyone expected the Yankees to sweep and move on to the World Series. Yet, with only one more win needed to clinch the play-offs, the Yankees lost it and the next three games in a row. The Red Sox advanced to their first World Series win in eighty-six years. By almost any metric, the Yankees choked under the pressure of their expected success.

Of course, no golf enthusiast can forget Greg Norman's exhibition at the 1996 Masters. His performance always makes the top-ten sport choke lists. After playing brilliantly in the first three days of the tournament, it seemed almost inevitable that Norman would take home the Green Jacket awarded to every Masters winner since 1949. Yet Norman's stellar play came abruptly to a halt that Sunday at Augusta. Even before the players got to the back nine, Norman had lost his huge lead to Nick Faldo and was visibly anxious. Everyone was in shock as Norman walked off the green five shots back and a second-place finisher. Some say he was almost in tears.

How could the opportunity for a win erode so fast? It was a question to which no one seemed to have an answer—including Norman himself. Speaking with *Golf Magazine* about the incident almost a year later, Norman said, "Never in my career have I experienced anything like what happened. . . . I was totally out of control. And I couldn't understand it."[1]

While we are on the subject of golf, it seems only fair to mention French golfer Jean Van de Velde, who is perhaps best known for his performance at the 1999 British Open. Van de Velde was so close to winning that his name was practically etched on the trophy, the Claret Jug. Arriving at the eighteenth hole three shots ahead after playing near perfect golf throughout the tournament, Van de Velde pulled—well—a "Van de Velde" when he took seven shots on the final par-four. Because of this, Van de Velde ended up in a three-way play-off that he eventually lost. But, it's not just the loss that people remember; it's also how it happened.

It all started with a mediocre drive on the eighteenth hole that Van de Velde tried to correct by going for the green on his second shot. Instead of hitting the green, however, the ball hit the grandstand running along the side and bounced backward into knee-deep grass. The next shot ended up in the creek. The following day sports pages across the world ran a picture of a dazed and confused Van de Velde, in the creek, with his socks and shoes off deciding whether to hit the ball out of the water. Van de Velde eventually took a penalty and left the ball where it was, but it was too late. His choke was already in the books.

When the pressure is on and everything is riding on the next move, sometimes even the most experienced athletes crack. Accomplished people in all fields have performed poorly when aiming for their best. Maybe that is why so many people felt for Norman during the Masters and Van de Velde during the British Open. They had the sympathies of everyone who has ever found himself or herself in their shoes.

At first glance, choking in a sport and other performance situations might seem to be outside the realm of what we can understand or predict. With new brain-imaging technology and the refinement of techniques for studying performance failure in the laboratory and on the playing field, however, psychologists have made some real inroads into why choking occurs and how to get rid of it.

Of course, it's often difficult for people to agree on what constitutes a choke in sports and other performance arenas in the first place. In my Human Performance Laboratory, recognizing poor performance under pressure is simple because we purposefully select folks so that they are all at the same high achievement level and then we dial up the stress to see when and why people's performance dives when the stakes are high. On the playing field, however, some athletes come to competition more skilled than others and what might seem like a choke by one player might simply be his or her opponent turning up the play. Indeed, in compiling a list of sport chokes, I had several arguments with close friends and colleagues about what should be included and what should be left off.

Choking is suboptimal performance, not just poor performance. It's a performance that is inferior to what you can do and have done in the past. We all have performance ups and downs, but choking occurs when performers perceive a situation to be highly stressful and, because of the stress, they screw up. Choking is most noticeable when an opportunity to win is squandered, perhaps because this is when the pressure to excel is at its highest. Choking is not random.

With the above definition in mind, poor performance under pressure can occur in individual and team play, when athletes compete directly against an opponent such as in tennis, or when they do not, as in golf. Admittedly, calling the choke in a direct competition situation can be tricky because you want to be sure that you are capturing a performance failure rather than a comeback by an opponent. Nonetheless, as any Yankee fan can attest, choking occurs in face-to-face play. Finally, choking can happen when athletes are in the midst of execution or before any movement has been made at all, such as Van de Velde's poor decision making on the last hole of the British Open. Yet hiccups during the unfolding of actions may occur for very different reasons than the factors underlying sloppy decision making when the stakes are high. That is why, in your own performance, you need to figure out what goes wrong and why so you can use the appropriate elixir to fix it. But first, some memorable sport chokes:

Greg Norman losing his lead and the tournament to Nick Faldo at the 1996 Masters.

Van de Velde's poor decision making on the eighteenth hole under pressure to win the 1999 British Open.

The Yankees blowing their 3-0 game lead to the Boston Red Sox in the 2004 American League Championship Series.

Lorena Ochoa at the 2005 U.S. Women's Open. "I thought I was going to win the tournament," Ochoa reported thinking before the eighteenth hole. Instead she hit a horrible first drive into the water and it was all downhill from there. Ochoa ended up finishing four shots back with a quadruple bogey on the last hole.

DePaul University's Skip Dillard was such a reliable free-throw shooter during the regular season that his teammates called him "Money." However, with twelve seconds left on the clock and a one-point lead against St. Joseph in the second round of the 1981 NCAA tournament, Dillard missed the first shot of a one-and-one opportunity. St. Joseph scored and won the game.

In the 1986 ice hockey playoffs between the Calgary Flames and the Edmonton Oilers, otherwise known as the Battle of Alberta, the Edmonton Oilers shot themselves in the foot (literally) when Steve Smith attempted a cross-ice pass near his own net. The pass struck his goaltender's skate and ended up in his own goal. The Oilers never recovered, losing the game 3-2, the series, and their chance for a repeat world championship.

England's penalty shootout performance in World Cup finals games might be the ultimate team choke. All you need to do is look at the statistics: Not counting the most recent World Cup, England has taken part in three penalty shoot-outs in World Cup final matches and lost all three. Fourteen attempts, seven makes, 50 percent success rate in a situation that is clearly in the shooter's favor.

Jana Novotna had one of the greatest breakdowns in Wimbledon history in the 1993 finals. Novotna lost the first set to Steffi Graf, 6-7, but won the second set, 6-1. In the third and final set, Novotna was one point away from going up 5-1, but instead she double-faulted and lost the game. She also lost the next four games in a row and the Grand Slam title.

Anyone who is an avid fan of the Olympic Winter Games will remember Lindsey Jacobellis's 2006 fall in the boardercross. Jacobellis was far ahead in the final race when she made a very poor decision. With all eyes on her, Jacobellis decided to be flashy on her last jump. She fell a few feet from the finish line, allowing Swiss snowboarder Tanja Frieden to take the Gold.

Michelle Kwan was a fan and judge favorite to win the gold in the 2002 Winter Olympics. Despite the fact that Kwan led after the short program, in her final long skate she was visibly stiff, two-footed a combination, and fell on her triple flip. Sarah Hughes took the gold.

Leading up to the 1992 Barcelona Olympics, Reebok ran commercials that depicted Dan O'Brien and his decathlete rival Dave Johnson competing for the gold. The competition never happened because, at the U.S. Team trials, Dan missed all of his pole vault attempts and never made it to Spain.

Alicia Sacramone's performance during the all-around competition at the 2008 Beijing Olympics. Despite being a team strong hold, when the pressure was on, Sacramone fell off the beam and performed miserably on the floor, taking the United States out of the running for the all-around Olympic gold.

On January 17, 2010, San Diego Chargers placekicker Nate Kaeding missed not one, not two, but three field-goal attempts in the 17-14 loss to the New York Jets in the AFC divisional play-offs. He missed a thirty-six-yard attempt in the first quarter wide left and a forty-yarder in the fourth quarter wide right. Kaeding hadn't missed a field goal from forty yards or closer all season. But in the divisional play-offs, when the pressure was on, his kicking went down the drain.

Kenny Perry had a chance to become the oldest winner of a golf major (at age forty-eight in 2009), but after playing seventy holes of the Masters almost flawlessly, he couldn't hold on to his lead. Everyone watched in disbelief as Perry bogeyed the final two holes and had to settle for runner-up. Perry seemed to be somewhat in shock as well. "I just didn't get the job done. . . . I'll look back the rest of my life saying what could have been," he said after the tournament.

• • •

Last December, my friend Mia invited me to spend a long weekend at her parents' new house in West Palm Beach, Florida. It was full-blown winter in Chicago so there was no way I was going to pass up an opportunity to spend a few days in the Florida sun. Mia's parents had recently retired and had traded in their typical workweek and long New York winter for year-round sun, sand, and golf at one of Florida's premier golf resorts.

Although spending your days on the links sounds like a welcome stress reliever from the grind of working life, things were not so straightforward, according to Mia's mom. One day over lunch, we got to talking and, to my surprise, I found out that retired life at the Ibis Golf and Country Club could be quite stressful. No one was playing the course for a Green Jacket, but just like Greg Norman at the Masters, these folks felt so much pressure to perform at their best that they were often performing at their worst instead.

When people move to a place like Ibis, they trade in their identities as impressive businessmen and women for new selves that are defined as much by their past successes as by their play on the golf course. This was certainly true for Mia's dad. He was no longer sweating about stock portfolios and mergers, but he was spending a lot of time worrying about how he would play on the front nine. Mia's mom had the extra complication of pressure from the hallowed Ibis Ladies Golf League. How low a lady shoots on the course can affect her social status in the community because the links are where dinner dates and party invitations are made, and the social scene is why people pay big bucks to live at Ibis in the first place. Mia's mom had already flubbed her first golf outing with the ladies and was nervously awaiting her next opportunity to play. She's actually a decent player, but just couldn't seem to pull it all together when it counted the most. She wanted to know why. What was going on?

In chapter 1, I related how a big sales pitch can go awry—especially when people are trying to deal on the fly with demanding questions from a client—and, in the past few chapters, I talked about how high-stakes testing situations can lead to poor performance when students need all their thinking and reasoning power to succeed. In short, we explored several situations in which working-memory is a must to perform well and we looked at how stressful situations can compromise our ability to use our cognitive horsepower when we need it most.

We talked not only about situations in which choking occurs because we don't have enough working-memory at our disposal; we also explored how to prevent and alleviate choking by clearing obstacles that impede the use of all of our thinking and reasoning skills. In this chapter we make a jump from mathletes to athletes and other performers. Choking in sports and in music have some similarities to choking on a big test in school, but there are differences, too. The goal is to give you techniques to alleviate choking that you can tailor to your own activity, whether stressing out before a major putt or nervously awaiting your solo on the stage.

WHY WE CHOKE ON THE GREEN . . . AND ELSEWHERE

People choke under pressure because they worry. They worry about the situation, its consequences, what others will think. They worry about what they will lose if they fail to succeed and whether they have the tools to make it. They may even conjure images in their head of the unwanted outcome—the flubbed performance, the missed shot, the fall on the ice.

Most professional athletes give the impression of being calm, cool, and collected during the many television interviews they do after a big game, so you might think that they would be good at controlling their thoughts and performance-related worries. Yet athletes *do* report thinking about the possibility of failure and sport scientists have shown that, when athletes think about themselves screwing up, they are more likely to do so. In golf, when negative self-talk pops into players' preperformance routines, simple putts go awry. In darts, when players imagine themselves missing the center of the target, they do. And in basketball, the more a player worries about missing an important free throw, the more likely he is to let his team down on the free-throw line in the final seconds of a big game. In short, the ability to control your thoughts and images during performance is crucial.[2]

When athletes think about themselves screwing up, they are more likely to do so.

Worrying (and trying to suppress your worries) uses up working-memory that could otherwise be used to maintain several pieces of information in mind at once so that you can make a reasoned pitch to a client or argue effectively with your spouse when you are pushing for a new kitchen remodel. Working-memory is also a must for a politician answering on-the-spot questions from a reporter. Our politician must harness all the cognitive horsepower at her disposal to ensure that her answers are accurate and not likely to ruffle any feathers. If not, the probability that the politician's remarks will be played over and over on twenty-four-hour news stations is quite high. Indeed, in many situations where people are faced with difficult thinking and reasoning tasks, worrying can harm performance by diverting brainpower.

But worries alone don't seem to be the cause of failure on the court or even on the stage—at least for performance greats. This is because brainpower is not necessarily what creates success in these arenas. Rather, the opposite seems to be true.

In basketball speak, *unconscious* is the term most often used to describe a shooter who can't seem to miss. And, according to San Antonio Spurs star Tim Duncan, "When you have to stop and think about things is when they go wrong." In dance, the great choreographer George Balanchine was known to urge his highly skilled dancers, "Don't think, just do." This recommendation is echoed on the links as well. As professional golfer Padraig Harrington has said, to hell with swing thoughts; he tries to keep his mind blank and think

> "Don't think, just do."
> —George Balanchine
>
> "Just do it."—Nike

only of the ball's flight path. Harrington, by the way, is known for his success on the PGA tour on Sundays, the final day of the tournament and arguably the most pressure-filled. And Olympic gold-medalist snowboarder Shaun White says that when he does one of his signature moves, a "frontside double cork 1080," which, in English, means two back flips and three 360-degree revolutions, "There's no time to think." "You let your body take over. I've done it for so long that when I'm not thinking, I'm riding the best." The key to high-level sports performance seems to be said best by Nike's motto: "Just do it."

Worries alone can't explain failure in sport because stellar performances in this arena don't rely heavily on the prefrontal cortex brain resources that worries co-opt. Not all activities use working-memory the way that computing demanding math problems or making reasoned arguments on the fly does. Complex motor skills, for example, are driven by procedural knowledge that resides in a network of brain regions, such as the basal ganglia and motor system, which largely circumvents the prefrontal cortex and the working-memory housed there. Even though worries seem to be related to failure in sport, they are probably not the direct cause of screwing up. So, what is?

The answer lies in what worrying catalyzes. When people are concerned about themselves and their performance, they tend to try to control their movements in order to ensure an optimal outcome. What results is that fluid performances—performances that run best largely outside of conscious awareness—are messed up. Now, if you are just learning the basics—first I step up to the ball, position my club head, my feet, etc.—this type of attention to detail can be a good thing. But if you are executing a skill you have done a thousand times in the past, the overattention that ensues when you are trying to perform at your best is exactly what makes you fall flat on your face.

Think about an action at which most people are experts—walking. If you are on your own hustling down the stairs, no problem. But if I ask you to think about how your knee is bending as you are shuffling down the stairs, you may have to slow down in order to reflect on the physical process, which may throw off your rhythm or throw off your step. You could end up stumbling on the stairs.

Let's take a step back and look at recent research on how experts' and novices' brains are organized to support athletic performance. This will give us a better idea of how athletes are able to perform at the top of their game and what factors cause them to crash and burn.

MIRRORS IN THE HEAD

When a football fan who once played the game is on the couch watching his favorite team at work—and doesn't have his hands around a beer can or a bag of chips—there is a good chance that he is actually mimicking some of the movements of the quarterback. Maybe he holds his hands as if he has the ball or even leans forward as the quarterback takes a step out of the pocket to run down the field. At first glance, you might think this is just overeager fan behavior, but recent research shows that this type of couch-potato mimicry is tied to the fan's once-high level of football skill.

When elite performers observe others in action, they don't just sit there idly and watch—at least this is true for their brains. Rather, these athletes seem to play off what they are watching inside their own heads. When expert ballet dancers watched videos of others dancing while their own brains were scanned using fMRI, the dancers' brain areas that are involved in producing the actions they were watching, such as the premotor and parietal cortex, were active. Interestingly, this activation pattern was only found when the experts watched dance moves from their own skill repertoire—moves that they, themselves, could perform. When experts viewed dancers performing moves they were unfamiliar with, these action brain areas were not as involved.[3]

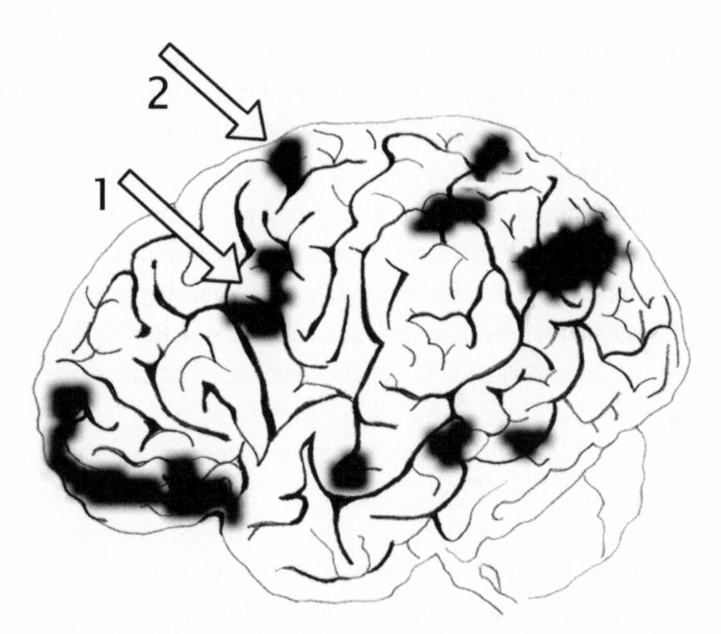

Areas of the brain that are more active when expert dancers watch their own dance style versus a different dance style. This picture depicts the left hemisphere of the brain. Active areas include (1) ventral premotor and (2) dorsal premotor cortex, brain areas that are involved in performing actions.[4]

When people watch activities for which they are highly skilled, they call upon not only brain areas specialized for seeing, but areas involved in action production as well. Brain networks involved in both doing and seeing have been termed *mirror networks*, and it's believed that highly skilled athletes rely on these types of mirrors in the head to predict the actions of others and to anticipate the outcomes of their own actions.[5] This is why skilled performers always seem as if they are two steps ahead of everyone else. In some ways, they are. Experts' brains have been rewired to play out actions in their heads before these actions have happened in reality.

Think about a professional snowboarder racing boardercross at the Olympics. This is the sport in which racers must compete against each other to be the first one down a course filled with tight turns and big jumps. If you are boarding in a crowd and you go off a jump, you have to know where you are going to land before you do so. This is the only way to prepare for your next move. This need for anticipation extends to most sports where you have to produce actions too quickly to get feedback from the environment. In soccer, goalies often have to start moving before the ball has left a kicker's foot, and in skiing, racers have to be thinking at least two gates ahead to have time to set up for their turns. It's as if experts' brains must predict into the future their actions before they have completed them so they can act fast and adjust when needed. Because of their immense amount of practice and experience, high-level athletes can take what they see or what they intend to do and form a good picture of how it will turn out.

Successful athletes don't reason step-by-step through their actions using their powerful prefrontal cortex as a guide. Rather, they are able to circumvent this step and, from what they see around them, start to play out what will happen in their head before it is actually realized. As a result, when elite athletes worry, it's not such a problem that working-memory is compromised, because these individuals don't rely heavily on cognitive horsepower to excel. Worrying often leads to problems, however. When you are at the top of your game and you suddenly find yourself worrying about screwing up, your desire to shine can prompt you to exert conscious control over what you are doing. This added control can backfire, disrupting well-learned sports and even musical performances that operate best outside the prefrontal cortex.

FRIENDLY FACES ARE NOT ALWAYS GOOD

Picture it: Mia's mom is on the green playing in the Ibis Ladies Golf League for the first time. It took her almost four months of schmoozing at the clubhouse to even get an invitation to play. She is at her first hole. To her joy, she made it to the flat, straight green under par and all she needs to get off to a good start is to sink a simple three-foot putt. This is a putt she has sunk so many times in practice she has lost count. This is a putt she knows, she understands, she can execute without a second thought. But, on this day, on this putt, she has other thoughts.

She thinks about how her social circle for the next several years may be dependent on her stroke count and she also notices that the other women in her foursome are watching her.

So, this putt is not exactly the same as the putts she has practiced for weeks. And when she steps up to the ball, performs her pre-shot routine, and executes her shot it becomes apparent just how different this putt actually is. Mia's mom starts thinking about herself, her swing, her stance, and as a result she does the unthinkable, given her ability—she misses the putt; she chokes under the pressure. The women watching Mia's mom weren't rooting for her to fail. They really wanted her to succeed. Mia's mom knew this, but their support didn't seem to help. In fact, it likely hurt.

For several years psychologists have known that putting a mirror in front of a person or videotaping him while he performs will make him more self-conscious—more aware of himself and his actions. This also occurs when we find ourselves having to perform in front of a live audience. Most interesting, the more supportive and friendly that audience is, the more self-aware we as performers get.

Yes it is quite satisfying to have your best performance witnessed by friends or family or a supportive audience in general. But keep in mind that it is also more painful to have your supporters see when you fall flat on your face. Of course, there are people who ham it up in front of an audience, who feed off their audiences' enthusiasm and get even better with adulation, such as the showboater who kicks it up a notch when people are watching. But, on average, the opposite seems to be true.

Think about pop singer turned country crooner Jessica Simpson. In 2006, Simpson was paying tribute to her idol and mentor Dolly Parton by singing her famous song "9 to 5" before a full house at the Kennedy Center, including President Bush, First Lady Laura Bush, and Parton herself. Simpson was so nervous she flubbed her lines and stopped singing all together. Performing in front of an idol and not wanting to fail is all it took for Simpson to flee the stage in tears. "Dolly, you make me so nervous. I can't even sing the words right," Simpson said of her flop on the stage.

Supportive audiences can increase your awareness of what you are doing and, if you are executing a highly practiced skill—say a three-foot putt on a straight green that you have made thousands of times in the past, or you are singing a song you have been rehearsing nonstop—this heightened attention to detail can actually mess you up. This was certainly true for Mia's mom playing with the Ibis ladies, but it also occurs in some unexpected situations.

In the mid-1980s, psychologist Roy Baumeister and his colleagues looked back at several years' worth of professional baseball (MLB) statistics from championship games in an attempt to understand how decisive games were won or lost in the most pressure-filled situations. What Baumeister found was quite surprising. When the home team was just one game away from winning the series, they won only about a third (38.5 percent) of the time.[6]

Heightened attention to detail can actually mess you up.

You might account for these statistics by suggesting that the away team obviously ratcheted up their play to prevent an early exit, but this isn't what Baumeister discovered. When he looked a bit further into the data, he found that the home team played poorer in these decisive situations; it wasn't the away team playing better. When the home team had a chance to win in baseball, they made more unforced errors. The 2004 New York Yankees' choke in the ALCS is a prime example of this. As I mentioned above, after being ahead in three games to none against the Boston Red Sox, the Yankees lost the next four games in a row—the last two at home.

Admittedly, Baumeister's data are somewhat counterintuitive—especially when you think about the commonly held notion that there is a home-court advantage. But, Baumeister thinks this advantage may disappear in critical matches when the pressures of performing in front of a home crowd, a supportive audience, cause players to try to control aspects of their performance in ways that disrupt it. More home-team unforced errors in baseball certainly supports Baumeister's view. Nonetheless, to quell the skeptics, Baumeister went into the laboratory for some added support.

In one experiment, Baumeister had people practice a video game called Sky Jinks.[7] The object is to steer an airplane with a joystick through an obstacle course as fast as possible. This game, similar to most sports skills, requires precise hand-eye coordination and quick reactions. Baumeister begin by giving everyone practice at the game to make sure they learned how to play so he could see what happened when he put the pressure on. Once his players were dominating Sky Jinks, he offered them cash if they could improve further. In addition, a stranger was also brought into the room who was scripted to act either obviously supportive, wanting the video game player to do well, or uninterested, showing no signs of caring how the Sky Jinks player fared.

The video game players were more prone to focus on themselves and their performance in front of a supportive audience, which led to a poorer performance than those playing in front of a neutral audience. Interestingly, in spite of performing more poorly in front of a supportive person than a neutral one, people felt that having a supportive audience was less stressful. The moral of the story: although we may think the support of others will always manifest itself as a home-court advantage, the opposite may actually be true—at least when the pressure is on.

PARALYSIS BY ANALYSIS

Why can being in front of a supportive audience cause people to choke? It comes back to the dangers of thinking too much. When we want to impress our friends, coaches, teammates, colleagues, or fans, we worry about doing so. To deal with these worries, we often attempt to take matters into our own hands. Our goal is to ensure success and so we start trying to control specific aspects of what we are doing.

A basketball player who makes 85 percent of his free throws in practice may miss the game-winning foul shot because, in an attempt to net the ball, he starts monitoring the angle of his wrist or the

Paralysis by analysis occurs when you attend too much to activities that normally operate outside conscious awareness.

release point of the ball. After thousands of free throws, these are not things that our basketball player would normally attend to and in trying to bring parts of his movement that normally run outside of working-memory back into it, he disrupts his performance. Our basketball player is demonstrating what we psychologists call *paralysis by analysis*.[8] Just as thinking about how and where we place our feet as we rush down the stairs may result in a fall, attending too much to activities that normally operate outside conscious awareness can lead to choking.

In a study conducted in my Human Performance Lab, we asked highly skilled college soccer players to dribble a soccer ball through a series of cones while paying attention to the side of their foot that was making contact with the ball.[9] This instruction was designed to draw attention to an aspect of their performance about which skilled players might not normally be conscious. We found that soccer dribbling was slower and more error-prone when the players paid attention to their foot in comparison to when they dribbled without any instructions. Paying attention to specific steps of what you are doing can be detrimental if, under normal conditions, these steps are not under your conscious control.

The same thing happens in baseball. When Division I intercollegiate baseball players were asked to take batting practice in a hitting simulator and at the same time attend to whether their bat was moving downward or upward over the course of their hitting movement, their performance suffered. Analysis of swing biomechanics revealed that the players' attention to their movements interfered with the timing of the different swing components.[10]

Attention can be counterproductive when it alters performance. Under pressure, people start worrying, which leads them to try to control their performance. Tasks that rely heavily on working-memory suffer from worrying. But sports skills and other activities that run largely outside working-memory are hurt, not because of worrying, but because of the attention and control that worrying produce.

DEGREES OF FREEDOM

When you first learn to perform a motor skill such as throwing a ball or hitting a hockey puck, there are innumerable possible ways you could coordinate the action because each joint involved—say, wrist, elbow, shoulder—has multiple possible movements or multiple *degrees of freedom* (df).[11] Because this is a lot of variability to handle, new learners tend to "freeze" some of their degrees of freedom by keeping joints rigidly locked in place or by coupling different joints together. Just think about how a kid learns to kick a soccer ball. Her leg is often straight, from the hip all the way through the knee to the ankle. Only with practice and instruction does she learn that effective kicks involve "unfreezing" the rigid links between the knee, hip, and ankle to allow for more flexible movement control.

Under pressure, we seem to regress and, in an attempt to exert control over what we are doing, we revert back to our beginner strategy of "freezing" different movement components. You can see this by looking at the movement patterns of weight lifters under training and competitive conditions. In situations where a lift is made in practice but not successfully completed during competition—that is, the lifter choked—the neck and hip joints of the lifter are often more tightly coupled than they should be. When the pressure is on, the lifter freezes his movements in a way that is counterproductive.[12]

Rock climbers who are high up on an indoor climbing wall and feeling anxious about it show a similar pattern of freezing under stress. Climbers exhibit more rigid and less fluent movements when they feel the pressure of being up high in contrast to when they are on the exact same wall, just at a lower level.[13]

This freezing pattern also occurs for musicians. Musicians' ability to let different joints of the fingers flow freely and independently when performing has been shown to be hindered when the stress of performance causes the coupling of movements that normally would not be as tightly linked together. As an example, the hands: when one is playing a complex piano solo involving different fingering movements of each hand, the hands must be able to operate independently. When stressed, people attempt to control execution in order to ensure success, but this can disrupt normally fluid movement patterns, making them more rigid, coupled together, and error-prone. The end result is choking under the pressure.

WHEN PAYING ATTENTION IS A GOOD THING

Although trying to control highly practiced motor skills can disrupt performance, sometimes attention to the details of your skill—or at least to the details of your surroundings—may be necessary. For instance, perhaps if Van de Velde at the 1999 British Open had devoted more brainpower to how he was approaching his second shot—should he go for the green or take it easy, given that he was so close to winning—his name would not be considered French for "choke." Or, if in the 2006 Winter Olympics, snowboarder Lindsey Jacobellis had been able to inhibit her tendency to showboat and had instead thought about the fact that her goal was to win at all costs, she would not have taken a careless jump and lost the gold. When people find themselves in a new situation—in the lead on the course or in the first-ever Olympic boardercross finals, devoting cognitive horsepower to their decisions and the potential outcomes these decisions may bring about can be a good thing.

Yet this is not always the case. When athletes are in familiar situations and have practiced their decision-making ability to perfection, careful consideration can get them in trouble even before they have made a move. Take the work of psychologists Joe Johnson and Markus Raab.[14] The researchers had team handball players watch videos of high-level games. At certain points the videos were paused and the players were asked to imagine that they were the person who currently possessed the ball and to select the best possible next move—such as whether to pass the ball or shoot—as quickly as possible. The screen then remained frozen and the players indicated all other possible actions that they, as the player with the ball, might perform. The researchers found that the first actions selected by the players were the best decisions (as judged by handball coaches). Also, the higher the skill level of the players, the more the first option was optimal.

You might predict that the final judgment would be a better decision than the initial one—after all, players had more time to make the latter choices. However, this was not the case. Because the players were so accustomed to being in the particular game situations that they were watching on the videos, they immediately knew what to do. This is precisely the reason that firefighters, police officers, and EMTs often practice making decisions in complicated situations—so that they are able to select the best possible option the fastest when a real problem arises. Nonetheless, when you find yourself in a novel situation, one that you have not encountered in the past, devoting working-memory to weighing out your options may be beneficial.

Another situation in which attention is important is when you want to make changes to your basketball shot or golf swing. You will need to unpack automated processes in order to alter your technique. Unfortunately, bringing skill processes that once operated outside conscious awareness back into working-memory does temporarily degrade performance—a necessity if you want to change in order to improve.

Professional golfers such as Tiger Woods and Nick Faldo faced performance slumps when they were making changes to their golf swings. It took Tiger at least a year to come back after he revamped his golf swing and Nick Faldo took the better part of three years to return to his previous level of performance after a swing change. Others have been less fortunate. In searching for more shot distance, the Australian golfer Ian Baker-Finch made changes to his swing and suffered a dramatic loss in form and confidence that eventually led him to withdraw from tournament golf. This was only five years after Baker-Finch had an impressive win at the British Open.

Unlike experts, novices need to think about what they are doing so they don't make beginner's mistakes. As a result, the performance of new players isn't hurt when they focus on their skill. In fact, beginners often improve with added skill control. When we asked beginning soccer players with only a few years' experience to dribble a ball as fast as they could through a series of cones, their performance actually got better when they were prompted to attend to the side of the foot that had just touched the ball. In contrast to experts, people just attempting to learn a new skill must devote attention to skill steps in order to ensure they develop.

Perhaps a study looking into the workings of the brain during beginning and skilled performance will help illustrate the distinction in how expert and novice performers control their skills. In 2005, psychologist Russell Poldrack and his colleagues watched people's brains as they developed from novice to skilled performers.[15] Unfortunately, it is impossible to swing or even to take a golf club inside an fMRI machine, because an MRI is a big magnet and a loose club would be drawn like a spear to the center of the magnetic field, beyond the control of the person inside the magnet. It's also difficult to get a good picture of what is going on inside people's heads when they are moving as they would have to be during a golf swing. So, Poldrack and his research group settled for a simpler motor skill to explore how the brain learns.

While people were lying in the fMRI scanner, they were presented with a symbol in one of four locations on a computer screen in front of them and asked to respond as quickly as possible to the symbol's location by pressing one of the four side-by-side buttons on a keyboard that was located closest to the symbol presented on the screen. As the study participants soon learned, the symbol's locations followed a consistent and repeating sequence, just like when people learn how to swing a golf club or dribble a soccer ball. The scan gave a picture of the brain as it first learns a sequence—showing the same inner process that, for instance, beginning soccer players undergo as they learn a dribbling sequence that involves kicking the ball, running, and kicking again. The scans also gave some insights into how the brain changes with practice.

Not surprisingly, as training progressed, people's ability to perform the sequence task got better and better. Poldrack found that when people performed the sequence before they'd had a lot of training, a wide network of brain regions was involved—especially prefrontal brain regions involved in working-memory and attentional control. After extensive training, however, activity in these regions decreased during sequence task performance. As you gain skills, the brain (at least the prefrontal cortex) often works less hard to support them.

These same brain changes are seen when people learn skills that require bimanual coordination (or coordination of both arms), which is so important for many sports and musical activities such as playing the piano or serving in tennis. Expert piano players use a smaller network of brain regions (less activation in frontal and parietal cortex) during the performance of a simple keyboard task compared to nonmusicians.[16] And when experienced golfers are putting or skilled marksmen are shooting, their brains work more efficiently than beginners' brains do, even though the experts and novices are performing the same task. In general, as learning progresses, there is a reduction in neural activity in the prefrontal cortex and other working-memory intensive brain areas (for example, the anterior cingulate cortex and parietal cortex) that were once needed to control step-by-step execution.

Sports scientist Bradley Hatfield at the University of Maryland has made his mark comparing the brain signals of expert athletes with those just starting to learn a skill, and has demonstrated that experts' brains are calm, cool, and collected in a way that novices' brains are not. Hatfield asks his athletes to wear a cumbersome cap full of electrodes that creates a picture of their neural activity while they play.

Billions of neurons in the human brain communicate by generating small electrochemical signals. When probes from an instrument that measures electrical energy are placed near a brain cell, a voltage change can be registered whenever the neuron is active. These electrical potentials are relatively small and cannot be monitored individually in humans without actually opening the head—at least not yet. Because there are so many neurons and because neighboring neurons frequently are active close together in time, however, the behavior of a group of neurons can be measured with probes placed on the scalp.

Hatfield has found that people who are at the top of their game differ markedly from the weekend warrior. Under normal conditions, Hatfield finds that the beginner's brain is firing all over the place. The experts, on the other hand, look calmer. In a study of expert and novice marksmen, Hatfield found that skilled rifle shooters exhibited less neural activity during the aiming period just prior to when they were about to pull the trigger. Even more telling was the *way* in which this activity changed for experts just before the shot. Skilled marksmen showed a decrease in the coherence or communication between motor areas of the brain and other brain areas like the prefrontal cortex (where reasoning and monitoring occurs). These areas stopped talking as frequently with each other right as the trigger was pulled. This decrease in coherence was not seen in the beginners.[17]

Reduced brain coherence can be good because it reduces the number of brain areas involved in an action or performance. "Too many cooks can spoil the soup" and too much brain interference with movement can make you choke. For highly skilled athletes, often times the movements necessary to carry out a successful performance are relatively preprogrammed, ready to go, and just need to be allowed to run off without much muddling and control. Hatfield and his colleagues think that a decrease in brain coherence right before the experts shoot indicates that they're "just doing it."

Under stress, of course, the expert brain can change. In a recent study, dart throwers, under the pressure of others watching them, showed an increase in the coherence between motor brain areas and those involved in attention, memory, and control.[18] That is, the brains of the expert dart throwers looked more like beginners' brains under pressure. This increase in brain traffic was accompanied by heightened levels of anxiety and less accurate dart throws. Similarly, when University of Maryland ROTC cadets took pistol shots during head-to-head competition, they showed an increase in brain coherence and a decrease in shooting performance compared to when the pressure was off.[19] As Hatfield puts it, excessive brain networking can result in undesirable alterations in the execution of highly practiced skills.

At least when it comes to training complex motor skills such as soccer dribbling or piano playing, practice alleviates some of the burden on the prefrontal cortex. Skilled performance becomes more efficient, more fluid, and in less need of constant attention and control. This is why too much muddling with your skill can be a bad thing. The best advice to prevent choking for skilled athletes and performers is to try to play "outside your head" or at least outside your prefrontal cortex. In the next chapter we will learn some tricks for doing this.

WHO CHOKES?

If you happen to be performing a complex thinking or reasoning problem that drains working-memory, then worries alone can lead you to choke. If you are performing a highly practiced motor skill, then worries in themselves do not lead to choking. But your attempts to consciously control your performance will trip you up.

We have already focused on why some people might be more prone to worry under stress than others. There are also differences in who is likely to overthink his performance when it counts the most. Several researchers have been exploring the difference in people's general level of self-consciousness (how aware people are of themselves) in order to get at who is most likely to choke when the stakes are high.

If you are highly self-conscious, for instance, then you would agree with statements such as "I'm aware of the way my mind works when I work through a problem" and "I'm concerned about what other people think of me." Some scientists speculate that people who are highly self-conscious should actually be less prone to choke under pressure, because they are used to performing under the type of hyperattention that pressure induces so stressful situations are not unusual for them. But the opposite prediction has also been made, that being highly self-conscious might make you very susceptible to thinking too much when the stress is on and, as a result, make you more likely to choke under pressure. Overall, the weight of evidence favors the latter prediction: highly self-conscious people are more prone to choke under pressure.[20]

As we saw in chapter 2, sport scientist Rich Masters and his colleagues devised a scale that assesses people's level of self-consciousness as a way to predict who is most likely to crumble under stress in athletic tasks. There is also a more general Reinvestment Scale,[21] as Masters calls it, that assesses people's level of self-consciousness in everyday situations. People answer questions with a yes/no, such as, "I reflect about myself a lot" and "I'm self conscious about the way I look." The higher a person scores on this scale, the more likely he or she is to choke under pressure in athletic, musical, and other high-stakes performance situations. The scale is below:

I remember things that upset me or make me angry for a long time afterward.	Yes / No
I get "worked up" just thinking about things that have upset me in the past.	Yes / No
I often find myself thinking over and over about things that have made me angry.	Yes / No
I think about ways of getting back at people who have made me angry long after the event has happened.	Yes / No
I never forget people making me angry or upset, even about small things.	Yes / No
When I am reminded of my past failures, I feel as if they are happening all over again.	Yes / No
I worry less about the future than most people I know.	Yes / No
I'm always trying to figure myself out.	Yes / No
I reflect about myself a lot.	Yes / No
I'm constantly examining my motives.	Yes / No

I sometimes have the feeling that I'm off somewhere watching myself.	Yes / No
I'm alert to changes in my mood.	Yes / No
I'm aware of the way my mind works when I work through a problem.	Yes / No
I'm concerned about my style of doing things.	Yes / No
I'm concerned about the way I present myself.	Yes / No
I'm self-conscious about the way I look.	Yes / No
I usually worry about making a good impression.	Yes / No
One of the last things I do before leaving my house is look in the mirror.	Yes / No
I'm concerned about what other people think of me.	Yes / No
Do you have trouble making up your mind?	Yes / No

Performers Barbra Streisand or Carly Simon, for example, are both known for being highly self-conscious about themselves and their performance and both have been known to choke onstage. "I think it's the whole experience of being in front of people, exposing yourself and your talent is so unnatural," Carly Simon has said. Indeed, in the 1980s, Simon was under so much performance pressure at a concert in Pittsburgh that she collapsed in front of ten thousand fans.

Being at the top of your game can also increase your chances of choking, according to Norwegian sport scientist Geir Jordet, who analyzed video images from all soccer penalty shootouts taken over the last twenty-five years from the World Cup (1982–2006), the European Championships (1974–2004), and the Union of European Football Associations (UEFA) Champions League (1996–2007).[22] In total, 366 kicks from 298 players were analyzed. Not surprisingly, at this high level, shots went in more often than not—overall about 74 percent of the shots were made. But Jordet found that the players who made the most shots were not always the ones you'd expect.

Those players who were the most publicly esteemed superstars—the recipients of prestigious soccer awards such as the FIFA, or South American player of the year—performed worse in major soccer penalty shootouts than players who had not yet received these major awards but would get them in the years to come. The current superstars scored only about 65 percent of their shots while the future superstar players were close to 90 percent. Jordet thinks this is because current superstars feel more pressure to perform at a high level than those who have not yet made it all the way to the top. Superstars have the love of the fans behind them and also the huge expectations that come with their success, which can make them more self-conscious. With that heightened self-consciousness, they also pay increased attention to their own performance, which, unfortunately, results in more shots wide of the goal.

You might be interested to know that the impact of stereotypes can play out in the sporting world as well. If you are negatively stereotyped about your ability on the court or field, then you are also likely to choke when under pressure. More about this below.

Out playing a quick nine holes at Ibis with her husband, Mia's mom was just about to drive off the first tee when her husband made a remark about men's and women's overall difference in strength. Women are just not built to have a lot of power, he said. Mia's mom shrugged off the remark and stepped up to the ball. Usually her long game was her strong point, but on the heels of her husband's casual remark, she had her worst initial drive in years.

Mia's mom had just experienced a stereotype threat, even though her husband was probably not trying to psych her out to impede her play. Also called trash talk, it is not so different in its effect from the more aggressive teasing that happens on the basketball court. Imagine a white guy shooting around one Saturday afternoon when a fellow teammate (who just happens to be black) jokes, "Everyone knows white men can't jump." The white guy would probably laugh off the remark and keep shooting, but it wouldn't be odd if he missed the next five shots in a row.

Can such minor comments really get under the skin of skilled players on the links and on the court? In other words, how could a comment that these players don't even buy into affect their performances in such a big way? Psychologist Jeff Stone at the University of Arizona has been asking this question for years and, in an effort to find an answer, set up a putting green in his laboratory where he invites golfers to see what they are made of. Sometimes Stone just lets them putt in peace. Other times he reminds them of stereotypes about how others expect them to perform.

In one study, Stone had black and white golfers take a series of golf putts on his indoor putting green.[23] Stone told some of the guys that they were taking putts as part of a test of "sports intelligence" because, naturally, putting is a game of thinking. He told other golfers, however, that the putts were part of a test of "natural athletic ability" because, as everyone knows, putting is all about hand-eye coordination.

Black golfers performed worse when they were told that putting measured sports intelligence rather than natural athletic ability. White golfers showed the exact opposite pattern—performing better when they thought the putts were a measure of sports intelligence. Merely framing a sports activity as diagnostic of negative racial stereotypes—African-Americans are not athletically intelligent, whites are not naturally athletic—was enough to push around golfers' putting success.

As in any high-stakes situation, when athletes are worried about perpetuating negative expectations, they tend to pay close attention to every step of their performance in an attempt to ensure success. As a result, "paralysis by analysis" ensues. Interestingly, getting golfers to play faster so they don't have as much time to think can counteract these effects. If you can also get the athletes to distract themselves—for example, by having them count backward by 3's as they putt—they perform better in the face of negative stereotypes.[24] Haste doesn't always make waste, and sometimes, distraction has its benefits. I'll talk more about techniques to counteract the pressure in the next chapter.

THE YIPS

Mia's dad likened his worries on the putting green to a bout of mental nausea. His mental game had really gone down the tubes since he had begun to play every day. Now sometimes he felt like he got a shaky feeling in his right arm when he knew he was about to miss a putt. He had what no one on the green or in the clubhouse ever wants to talk about: Mia's dad had a case of the yips.

Most of us have heard the term *yips* in sports before, but what are they exactly? At the most basic level, the yips are involuntary jerks, tremors, and spasms in the extremities that disrupt the execution of fine motor skills. Usually talked about in relation to golf putting, the yips occur in other sports, too. In archery, the yips are known by a different name: target panic or target fever. But as in golf, the affliction is so feared that players often refuse to say its name out loud. The yips may also be at the center of what former New York Yankees second baseman Chuck Knoblauch faced when, all of a sudden, he discovered he could no longer throw to first base. Some even speculate that Shaquille O'Neal's dismal free-throw performance is a case of the yips. The yips have even been known to creep in and affect nonsports tasks, too.

The yips afflict people performing a variety of professions, including musicians, stenographers, dentists, and surgeons. They are most likely to hit folks who overuse the muscles involved in guiding precision movements. So it's no surprise that as Mia's dad ups his time on the green, his likelihood of experiencing the yips increases as well. Initially, the yips was thought to be a purely psychological phenomenon related to feelings of anxiety and stress, but researchers now believe that the yips can have a physical cause, too. Indeed, the yips are now often described as a form of dystonia, a neurological disorder typified by involuntary movements that result in a twisting and spasming of body parts. Dystonia can occur in a single muscle group (focal dystonia) or in a more generalized fashion that affects multiple body parts (generalized dystonia). Many dystonias are isolated to specific tasks or situations in which a person is required to perform a well-learned and repetitive movement (such as a golf swing). The cause of dystonia has been linked to abnormalities in the functioning of the basal ganglia and motor pathways of the brain, as well as to head injuries and stroke.

After doing research on dozens of golfers who have manifested cases of the yips, Aynsley Smith and her colleagues at the Mayo Sports Medicine Center in Minnesota came to the conclusion that there are actually two types of yips.[25] Type I yips are a form of focal dystonia. Type II yips result from performance worries that increase self-awareness and attention to performance: paralysis by analysis. Although Type I and Type II yips are characterized by different processes—the former starting from the motor pathways in the brain and the latter from prefrontal regions that are too active in trying to control what people do—all forms of the yips are thought to manifest themselves in similar ways. In the case of golf putting, this amounts to a jerk, twitch, tremor, or freezing of the putting stroke that ultimately disrupts play. Smith has likened it to a hiccup in the wrist.

The yips have afflicted professional golfers like Ben Hogan and Bernard Langer. In fact, a golf grip has even been named after Langer in reference to his attempts to beat the yips by changing up his grip. In a standard golf grip, the right hand is placed below the left hand and the thumbs are vertically aligned. When golfers turn to the Langer grip, the left hand is placed low on the putter shaft and the shaft is aligned along the left forearm. The idea behind the change is that, because a golfer has practiced for so many years with his normal putting grip, his brain may not only be programmed to execute his putt in a certain way, it may also be programmed to yip. These days there are no rules on putting grips, so an overhaul may do the trick in getting rid of the yips. The hope is that the change-up gets your brain to reprogram the circuits that you need in order to execute your shot so that it clears your brain and body of the yips. You get a clean slate.

It worked for Langer and for golfer Sam Snead as well. Late in his career, Snead developed the yips. To deal with it, he started putting in what is termed a sidesaddle position. Snead relearned his putt without relearning the yips. But not everyone is lucky enough to beat the yips with a mere grip change. In fact, many golfers leave the game because of it. Mia's dad has recently started using a cross-handed grip (where he places his left hand below his right hand) combined with a longer putter, which seems to be working for him. But he admits that he is worried about the yips returning. Unfortunately, worrying about it may be the worst thing he could do.

WHERE ARE WE NOW?

Some say that the free-throw line is the toughest place on the court for a basketball player to remain calm. Although it's only fifteen feet away from the basket, it can seem like a football field in distance. The free-throw line is where games can be won or lost by one player or one shot. It is where the pressure is on.

Recently, a group of psychologists at the University of Texas at Austin obtained game transcripts from every NBA regular season and play-off game from the 2003–2004, 2004–2005, and 2005–2006 seasons, so that they could examine players' free-throw shots in the final minutes of the game. They found that players tended to shoot worse than their career free-throw shooting average when their team was behind by only a point—when the pressure was at its highest because a make or a miss could really shift the outcome of the game.[26]

Free throws happen in pressure situations and coaches don't usually have their players practice these shots under stress. The game-winning free throw simply can't be mimicked in practice, where a miss means nothing and when fans, scouts, or sportswriters are not around to notice. When the ability to keep your team in the game rests entirely on your next shot and you haven't had an opportunity to get used to this pressure, your performance—even if you're a superstar athlete—can suffer.

New research suggests, however, that if coaches have their players practice under the stress of competition, actual game play looks different. Of course, you can never mimic in practice the exact same level of pressure you feel in a game-winning situation. Nonetheless, psychologists have developed a number of practice and performance techniques that help thwart the negative effects of performing under stress. In the next chapter, we explore the best ways to ensure optimal performance when the stakes are high on the court, on the green, and even on the stage.

CHAPTER EIGHT

FIXING THE CRACKS IN SPORT AND OTHER FIELDS

ANTI-CHOKE TECHNIQUES

My office is never a good place to get work done. Every time I sit down to read about new research going on in my field or to look through the results from one of our latest projects, someone inevitably arrives at my door. It might be a University of Chicago undergraduate looking for help with a term paper, a colleague needing a quick answer to a question, or a Ph.D. student asking me to look over his research talk for an upcoming conference.

The phone is another story altogether, especially on Monday mornings, when I often get phone calls from parents hunting for a psychologist to work with the aspiring athlete in their family—an athlete who didn't perform up to his or her potential over the previous weekend. Frequently these kids are high-caliber players, at the top of their age group and vying for titles and championships, but when the pressure is on, things don't always go well. Interestingly, these signs of cracking under stress often start to emerge right around the time that parents, coaches, and the kids themselves realize the potential for college scholarships, the Olympics, and other endeavors. Before then, these kids' sport dominance came pretty easily. Now their parents want to know what they can do to get it back.

This certainly described David's situation. Early one spring morning I got a call from his dad. A freshman at New Trier High School in Winnetka, a suburb of Chicago, David had been playing soccer since his parents had given him a dime-store plastic ball at age five and he had been excelling on the field ever since. Now, as a high school goalkeeper, David was starting to catch the eye of several college coaches with his level of play.

David's dad had to admit that he had always been a little surprised that his son never really seemed to mind the pressure of being the last defender before the goal. But all that had changed this season when David was put in the unheard-of position of being a freshman starter on the varsity team. David played a good season, but in a few games he had made a mistake or two that had probably cost New Trier the matches. Now his dad could see that the pressure was taking a toll.

New Trier had played in the state championship game the previous weekend. Regular game play had ended in a 1-1 tie and overtime didn't change the draw so the game went into penalty kicks. The standard procedure in soccer when two teams take penalty kicks is for each team to take five shots on goal. The team that makes more, wins. Penalty kicks are the ultimate one-on-one situation. The shooter faces the goalkeeper and they're the only ones in play. To be fair, the odds are definitely in the shooter's favor. But, the lore goes, if a goalkeeper can get his or her hands on the ball, he or she should be able to prevent it from going in.

The first shot David had to defend was a good kick to the upper right corner—nothing anyone could have done. But in the next four, two of them went right through his hands. He had really botched the penalty round and New Trier had lost the state title to Edwardsville because of it. David's dad had never seen his son play so poorly when it counted the most. After the game, all David could say was, "I choked"—hence the phone call.

Over the course of his adolescent life, David has spent thousands of hours on the field, in the weight room, and running stairs with private coaches, and in most game situations it was easy to see the product of David's hard work. But none of this physical training seemed to help when the pressure was at a maximum in the state championship penalty shot round. David's dad wanted to know whether I could help train his son's mind as others had trained his body. If David could learn to deal with pressure when the stakes were highest, then future Illinois state titles, college scholarships, and perhaps even Olympic medals wouldn't slip—like the penalty kicks—through his son's hands.

High-pressure environments induce a variety of brain and body reactions. Your heart rate goes up, your adrenaline kicks in, and your mind starts to race—often with worries. When the worries begin, many people do something that seems quite logical on the surface: they try to control their performance and force an optimal outcome. Unfortunately, this increased control can backfire—especially for well-learned skills—because bringing your conscious awareness to skills that once operated outside your working-memory and prefrontal cortex can disrupt them. You cause yourself paralysis by analysis and you choke under pressure.

Over the past several years, psychologists have been using newly discovered information about how the brain supports exceptional performance in order to develop new training regimens, practice schedules, and performance strategies to alleviate poor performance under stress. Let's take a look.

In January 2009, the National Collegiate Athletic Association (NCAA) adjusted the grade in which a basketball player could be considered a college prospect from the beginning of high school (ninth grade) to the beginning of junior high (seventh grade). The change was designed to close a loophole in the college recruiting rules that allowed college coaches to bring junior high school players to private camps at their universities. This type of recruiting is not allowed at the high school level, in order to prevent the unfair and pressured steering of high school kids to particular universities. So college coaches were attempting to get a recruiting edge by inviting promising junior high school players in for summer camps instead. With the new rule change, such practices are no longer allowed. Now these junior high school players, like their high school counterparts, are considered recruits.

At first glance, the NCAA rule change seemed nothing but positive in preventing the unfair cajoling of seventh- and eighth-grade players by eager college coaches. But a downside to the rule change is only now coming to the forefront. Simply put, treating seventh- and eighth-grade students like their high school counterparts clearly identifies them as prospects for college ball. As a result, now junior high students across the country are being ranked for their ability on the court.

When seventh graders are recognized as a college prospect, their parents often hire trainers and coaches and end up paying big bucks for basketball camps so their kids can develop techniques that may one day help them "own the court." However, these regimens often neglect to train the children's minds for the pressures that lie ahead.

Getting marked as a future college phenom may empower a young player to begin to pursue a basketball career—it may instill the confidence and drive he needs to succeed in today's competitive sports environment. Yet being identified in this way can also put pressure on a player because everyone expects him or her to be the best. In order to handle that pressure—when college coaches are watching, NBA scouts are in the stands, and the other team is trying to shut you down—players need a particular set of practice and performance strategies. This chapter uncovers these anti-choke techniques.

GET USED TO IT!

While there is certainly merit to the age-old adage "practice makes perfect," practice has a better chance of producing perfection when players practice under the conditions they are actually going to face in competition. It's difficult to mimic the pressures athletes will face in the potentially game-winning free throw or penalty shootout or that a performer will face during his solo on the stage, but as we have seen throughout this book, training under even mild levels of stress can help.

Think about one of the most stress-filled situations a basketball player faces on the court: a trip to the free-throw line. Free throws are interesting because they are a relatively easy shot—no one is guarding you and the basket is a mere fifteen feet away. Yet players often miss. This is true even at the college and professional level. In 2009, free-throw shooting success in men's college hoops was about 70 percent. The NBA statistics don't look much different, hovering around 75 percent for the better part of fifty years. One reason for the lackluster free-throw performance of even the most skilled players is that these shots are often taken in high-stakes situations—to win a game or at least to help one's team get closer to clinching the lead. Even more telling about players' dismal free-throw success is that most teams don't schedule free-throw practice. Even players who do practice free throws don't do so under conditions that mimic the pressure-filled situations they face in game play.

It's not unheard-of for basketball coaches to have their players casually take free throws at the end of practice or to expect players to practice them on their own time, since they don't really need anyone around to play against. But if you have to make two free throws to win the game and you haven't practiced shooting them under stress, being up at the line with everything riding on your next shot may feel completely foreign.

Some coaches who have their players take free throws when the stress is on have had some remarkable success. Take Southern Utah University's coach, Roger Reid. In the middle of a practice scrimmage when the players are least expecting it, Coach Reid will halt everything and immediately send his players to the free-throw line. A made shot and the player gets to catch his breath. A miss means a sprint around the court. When Reid arrived at Southern Utah, he inherited a team that ranked 217th in free-throw percentage. As of 2009, they were ranked number one and are shooting just above 80 percent.[1]

Even though having to sprint for a missed shot doesn't create a huge amount of pressure, mild stress training is beneficial because it gets players used to performing with something on the line. You see the beneficial results of training under the gun in all sorts of sports. For instance, at the English Institute of Sports, performance psychologist Pete Lindsay has been working with British Judo's development athletes during training camps to help them up the pressure during practice so the athletes are used to it once they get to the real competition.[2]

To do this, the judo athletes might be told they have to fight in a smaller area of the mat during training or the coaches attempt to stress the players out by having them perform demanding combat drills or exercises before they step in for a practice match.

Our goal is to "reduce the gap between training and competition," says Lindsay. And it seems to be working. The judo athletes experienced pressure training before the team competed in the Junior European and Junior World Championships. The squad won three medals at the Junior Europeans and had the most successful Junior World Championships in fourteen years, winning two bronze and a silver medal.

Think about the common technique of exposure therapy, used to help people deal with their fears and anxiety about, for instance, spiders or heights. Exposure therapy involves habituating individuals to what they fear most so that, over time, people can learn to reduce the negative associations with a particular trigger and function better. For instance, people who are terrified of heights and receive height exposure therapy, including computer-simulated rides in a mock elevator, cope with their anxieties better than those who do not. These height-phobics are more likely to ride in elevators at a later date. Riding in an elevator over and over makes heights less frightening. Similarly, training in stressful situations minimizes the possibility of the choke as you gradually become accustomed to the pressure so that situations that once made you uneasy no longer feel unusual or threatening.

The more people practice under pressure, the less likely they will be to react negatively when the stress is on. This certainly seems to be true for professional golfers like Tiger Woods. To help Woods learn to block out distractions during critical times on the course, his father, Earl Woods, would drop golf bags, roll balls across Tiger's line of sight, and jingle change in his pocket. Getting Woods used to performing under stress helped him learn to focus and excel on the green. Of course, no training method is foolproof and even the most accomplished athletes may falter under pressure sometimes. When the weight of personal blunders off the course gets too heavy, even your high-level performance on the links can suffer.

Riding in an elevator over and over makes heights less frightening. Similarly, training in stressful situations minimizes the possibility of the choke as you gradually become accustomed to the pressure.

Training under pressure can do more than just get you used to stress. It can also help you get accustomed to the overattention to performance that often accompanies high-stress situations. Acclimatizing performers to conditions that heighten self-awareness during training adapts them to performing in that state. When golfers practice putting while being videotaped and are told that the videotape will later be watched by golf coaches, they perform better under a subsequent pressure-filled situation than those who did not receive this acclimatization training.[3] This is true in the musical world, too. When musicians were asked to give a performance in front of an audience—some of whom were said to be rating their play—those who had practiced under the watchful gaze of a video camera beforehand performed much better than those who had practiced in isolation.[4]

Giving a performance in front of people who are judging you or even people who want you to succeed is not unusual in the sports and musical worlds. Performers who get used to this type of audience scrutiny to begin with, for example, by being videotaped during practice and thinking that others will review their practice, reduce the self-conscious attention to detail that they might otherwise feel in the actual recital or game.

Young basketball prospects can benefit from this type of training, especially if they want to make it in the college hoops scene. It is something that our soccer goalkeeper, David, will need if he is going to continue to catch the eye of college coaches. At the boys state soccer final, in the penalty shot round, all eyes were on David. If he had spent some time in practice getting used to defending penalty shots under the scrutiny of a full stadium or at least with some semblance of an audience present, perhaps he would have been less likely to choke under the pressure.

Keep in mind that, while practice under pressure is important, so is what you practice. Between 1980 and 1990, there were over 170 separate incidents of mechanical malfunctions in U.S. Air Force fighter aircraft, ranging from loss of an engine to landing gear malfunction to failure of multiple systems or controls at once.[5] Researchers looked at how pilots dealt with mechanical mishaps that are explicitly practiced in the flight simulators and are known as "fair game" for flight-performance evaluations versus malfunctions that are not normally practiced. Just to give you some idea of what is practiced and what is not, pilots routinely practice what to do if they lose an engine, but it is hard to practice all the particular combinations of multiple control malfunctions at once because they don't know what they might be. Talk about the ultimate pressure situation: the right move when flying means survival and the wrong move can mean sure death.

While practice under pressure is important, so is what you practice.

As you might expect, the pilots' number of flying hours was a good predictor of effective performance and decision making under the gun for the malfunctions that were practiced in the past. The malfunctions that were never practiced, however, were not dealt with as effectively and this was true for both veteran flyers and pilots who were newer to the skies.

Having experienced pilots in planes when there is a malfunction is a good thing. It's even better, however, if the pilot has practiced what to do when the emergency occurs. As an example, take US Airways flight 1549, which left New York's La Guardia Airport on January 15, 2009. When geese were sucked into the jet's engines, veteran pilot Chesley B. "Sully" Sullenberger was forced to make an emergency landing in the Hudson River. Although the particulars of geese obstructing an engine were likely not in this pilot's practice repertoire, the more general case of a dead engine was and Sullenberger knew what to do. Sullenberger had been a pilot with US Airways since 1980 and had spent seven years in the U.S. Air Force before becoming a commercial pilot. When the engines died, Sullenberger went through well-practiced procedures and checklists and was able to turn on thrusters to land safely in the Hudson River. There were only minor injuries and no deaths.

Military and commercial aviation relies heavily on simulators to train pilots. This is true in other performance situations as well. Surgeons practice on operating-room simulators and firefighters run practice drills constantly to keep themselves at the top of their game. It seems clear that practice is beneficial, and that the closer your practice can mimic what you are likely to face in the real do-or-die situation, the better you will be able to handle unexpected occurrences in times of crisis.

JUST DO IT!

At the 2008 Beijing Olympic Games, U.S. gymnast Alicia Sacramone stepped onto the mat to compete on the beam. It was the all-around team finals, and the U.S. and Chinese teams were vying for the gold. The pressure was on for Sacramone to pull off a perfect performance, but she was ready to go. The judges were not, however, and Alicia was told to hold back, twice, before she was eventually given the green light to mount the beam. Unfortunately, when it was finally Alicia's turn to shine, she didn't. She fell off her initial jump onto the beam and had to start from the floor, with deductions that would help prevent her team from winning the gold.

Then it was on to the floor exercise. Despite Alicia's stumble on the beam, the U.S. team had pulled within one point of the Chinese team with only the floor exercise remaining. Sacramone had a chance to turn everything around and come through in a pinch. But instead of producing her best success, Alicia stepped out of bounds and landed on her back. America's hope for the gold was over. "It's kinda hard not to blame myself," Alicia Sacramone said after the meet, when the Americans walked away in second place.

Before her beam routine, Alicia Sacramone was a team leader, the face of optimism, hope, and inspiration. But after her fall off the beam, her attitude and performance the rest of the night seemed to fall, too. We will never know for sure what the catalyst for Alicia's initial screw-up was, but one likely candidate was the fact that she was forced to pause (twice) before she even started her beam routine.

"They put her name up with a stop sign," U.S. coach Martha Karolyi said. "She couldn't go once, she couldn't go twice . . ." Sacramone was iced.

Thinking too much can get in the way of what you are doing—especially if you are performing a skill that you have practiced to perfection. When people have a lot of time on their hands to prepare for a kick, a shot, or a solo, this ample preparation time can hurt their performance. Although we have always heard that "haste makes waste," sometimes haste actually prevents waste.

Think about American skier Julia Mancuso at the 2010 Winter Olympics. Halfway down the hill during the women's giant slalom, Mancuso was "yellow flagged" and had to abort her run because teammate Lindsey Vonn had crashed right before her and was not safely off the course. Mancuso had to start the race over and, as if that weren't bad enough, there were arguments about how to get her back to the start gate (snowmobile or gondola). Mancuso's ascent back to the start house was actually stalled when an Olympic official stopped the snowmobile she was riding on and demanded she take the gondola to the top. Mancuso ignored the mandate, but the damage was done. Mancuso's rerun left her in a disappointing eighteenth place and the defending 2006 Olympic gold medal champion didn't even make it onto the podium in 2010. "I was emotionally exhausted," Mancuso said after her first run. Stalling before you even begin, or having to restart your performance once you have started, can have dire consequences.

More time means added opportunities for counterproductive control to take over skilled movements that are best left alone. fMRI research shows that when people attend to their movements after they have learned them to perfection, the prefrontal cortex—the center of working-memory—becomes highly activated.[6] Because the prefrontal cortex doesn't do the lion's share of the work in controlling highly skilled movements under normal conditions, it is likely the seed of where counterproductive attentional control begins with added time. There are countermeasures you can take to prevent icing, however, which we'll discuss.

Not long ago, I spent the afternoon with David Rath and Chris Fagan, two coaches of the Hawthorn Hawks, an Australian rules football team. The sport is also referred to as football, footy, or Aussie rules. On the surface, Aussie rules looks quite different from American football, but it actually shares a number of similarities, which is why the coaches were in the United States to talk to NFL coaches about football training. They were also visiting with a few psychologists around the country in the hopes of picking up psychological techniques to help their players function at their best in the most pressure-filled matches. The Hawks had just finished their Australian Football League season in first place and the coaches were looking for ways to keep their team at the top.

While we were having coffee, coaches Rath and Fagan told me a story about one of their star players, Lance Franklin, who had had some problem with his free kicks. Although Franklin was regarded as one of the team's (and league's) best kickers, Franklin had struggled earlier in the season.

In Aussie rules, a player gets a free kick when he has been fouled. As Rath and Fagan told me, there is often a delay between when a penalty is called and when the free kick is initiated and this delay was proving detrimental to Franklin's kicking success. Interestingly, Franklin had figured out a trick for dealing with it that seemed to be working. Instead of stopping before the kick, as is often the norm, Franklin would just keep running, roll around, and take the kick as quickly as he could.

The coaches and I agreed that Franklin's "just-do-it" strategy was a good one—especially if it prevented the type of mistakes he had made in the past. I even pointed out that other athletes have similar strategies for managing their performance time. Golfer Aaron Baddeley, for example, has a relatively short pre-putt routine (a four-count from the moment he grounds the putter to the moment he strikes the ball) and Baddeley is consistently rated as one of the best putters on the PGA tour. In my Human Performance Laboratory, we have shown that skilled golfers putt better when we instruct them to putt as quickly as possible. Of course, new golfers need plenty of time to think about what they want to do, because attending to the step-by-step details is important when you are just learning the tools of your trade. But once a skill is well practiced, too much time—allowing for too much attention to detail—can be a bad thing. Indeed, in American football, a slowing-down tactic is often used to disrupt kicking performance.

In American football, defending teams use a tactic to "ice the kicker" to disrupt field goal kicks just prior to the snap. Coach Karolyi thinks that the Chinese tried to do the same with Alicia Sacramone at the Olympics. The general lore is that icing gives a performer time to worry about screwing up. My work suggests it also gives athletes too much time to think about what she or he is about to do. Statisticians Scott Berry and Craig Wood also think that icing works—especially when the stress is on.[7] They looked at field goal kicks in the 2002 and 2003 NFL seasons, concentrating on kicks in which a make would give the kicking team a lead in the game or at least a tie with less than three minutes to go or in overtime. In other words, the researchers focused on kicks where the pressure to get the ball through the uprights was pretty high. They found that calling a time-out before the kick did in fact work.

As the researchers themselves reported, the drop in success when a time-out is called "implies that a psychological effect of pressure exists, and is compounded by more time to dwell on the kick." If pressure-filled situations prompt performers to try to control execution in a way that alters their normal routine, then extending the time that kickers have to scrutinize their shots will indeed exacerbate the likelihood of choking.

Recently, a new tactic has come into vogue in the NFL, called Icing 2.0. In response to a rule change that allows time-outs to be requested from the sidelines, coaches have been calling time-outs right before the ball is snapped. The end result is that the kicking team doesn't realize the kick didn't count until after the play is over and they have to take the kick again. The rule change was enacted in 2004 in order to allow coaches to call sideline time-outs so they didn't have to waste precious seconds relaying the time-out call to the player on the field. But it is now being used to make kickers kick again.

Icing 2.0 was put on the map by Denver Broncos coach Mike Shanahan in the second week of the 2007 NFL season. As Oakland Raiders kicker Sebastian Janikowski lined up for a game-winning field goal attempt, Coach Shanahan walked up to the official and called time-out just as the kicker looked up for the snap of the ball. Janikowski celebrated his fifty-two-yard coup right up until the officials announced that he would have to take the kick again. The rekick hit one of the field goal posts and missed. The Broncos went on to win the game.

It's too early to tell whether this new tactic is effective. Indeed, before Icing 2.0 can really be evaluated, the NFL might change its rule to prevent it. But one thing is for sure: altering performance conditions so that athletes have to change how they execute their well-practiced skills can be detrimental.

DISTRACT YOURSELF

As I was on the phone talking to soccer player David's dad, I couldn't help thinking that a timing issue may have contributed to David's inability to shine in the goal. Waiting for a player to take a shot can leave a goalkeeper with a lot of time on his hands. If David started thinking about how he was standing, which foot he might push off of as the ball was shot, or even how he was going to try to deflect the ball, this may very well have disrupted his ability to operate at his best. After all, David had been saving penalty kicks since he was five years old. Of course, in David's position, he can't do much about time because it's the kicker, not the goalkeeper, who gets to decide when the play commences. If the kicker drags out his shot and David starts overthinking his next move, what should he do?

You might assume a good option would be just to tell the goalie not to overanalyze his save, but research shows that telling people not to think about something isn't effective at suppressing unwanted thoughts or disrupting an inappropriate focus of attention. In fact, when people are told not to think of something—the famous case is a white bear—they tend to do it more.[8] Fortunately, a number of active focusing strategies have proven beneficial for limiting players' tendency to overthink their next move.

Players might try to focus, for instance, on the dimple pattern or on the manufacturer's name on a golf ball or the seams of a rugby ball. Rugby union goal kicker Jonny Wilkinson focuses intently on the precise point of the ball that he wants to strike. He combines this focus on the ball with a focus on a particular person in the crowd who just happens to be sitting between the posts he is trying to kick through. In pistol shooting, a rifleman focuses on the target he is trying to hit rather than his trigger finger. In opera, a singer might focus on the melody of the song rather than how exactly she is going to hit the high note she has hit several times in the past. And, in microsurgery, a doctor might focus on the artery he or she is trying to repair rather than the precise movement of his or her fingers and hands.

Tim Gallwey, coach and author of the Inner Game series of books, talks of this focusing process in tennis when he encourages players to say "bounce" at precisely the moment the ball lands on the tennis court and "hit" at the moment it makes contact with their racquet. Focusing on something other than the movements of your performance can train the brain to stay out of what you are doing.

Focus cues have been used to cure cases of the yips in golf. One near-scratch golfer had difficulty initiating his downswing. It had gotten so bad that he would take numerous false starts on the course in which he froze at the top of his swing. The intervention that proved most effective was to have the golfer focus on a three-syllable word to match the rhythm of his swing. He used the song title "Edelweiss" from *The Sound of Music.* The golfer said "Ed" to correspond to initiation of his back swing, "el" to match the top of his back swing, and "weiss" at the point of contact with the ball.

Focus cues can also be beneficial in times of stress. My colleagues and I have found that skilled golfers make more putts in do-or-die situations when we distract them from what they are doing than when we leave them alone. In one study, we had golfers listen to a series of words being played from a loudspeaker while they putted. Every time the golfers heard a word, they had to repeat it out loud. The process of drawing golfers' attention away from their own performance improved their putting under pressure.[9]

A recent study of basketball players echoes our findings. Australian researchers found that players with a propensity to choke made more free throws under pressure when they listened to music ("Always Look on the Bright Side of Life" from Monty Python's *Life of Brian*) than when they did not. Distracting these athletes from focusing in too much detail on their shot allowed them to execute their free throws with minimal involvement from working-memory and the prefrontal cortex. Because this involvement often slows down fluid movements and creates new opportunities for error, the players sank more shots when they were distracted.[10]

FOCUS ON THE GOAL

Like many sport skills, saving a one-on-one shot on goal in soccer requires technical precision as well as effective decision making. The goaltender has to decide if he is going to try to anticipate where the

Focusing on what to do (a strategy focus) rather than how to do it (a technique focus) can help prevent cracking under stress.

shooter is going or whether he is going to just wait and react to the shot. The shooter has to determine what kind of shot he is going to take in the first place—will it be a shot to the upper right corner or a line drive on the ground? In this type of penalty situation, choking (on either the part of the kicker or the goalkeeper) often results from too much attention to the details of what they are doing. As a result, focusing on what to do (a strategy focus) rather than how to do it (a technique focus) can help prevent cracking under stress.

Recently, sports scientist Robin Jackson conducted a study with soccer players where he showed the benefits of a strategy focus.[11] Jackson is on faculty at Brunel University in London and, like most Englishmen, is a big soccer fan. Indeed, when he is not studying soccer players for his job, Robin can often be found watching local matches or kicking around the ball with his kids in the backyard.

Before having highly skilled soccer players dribble a soccer ball through a series of cones on the soccer field, Jackson asked them to set some goals for themselves. The players had to choose a goal that they felt would help maximize their dribbling success. Some of the players set goals directly related to the movements or technique required to dribble (for example, "Keep loose with knees bent"). Others set goals that related more to strategic elements of the task (such as "Keep the ball close to the cones").

Jackson found that those players who set technique goals performed worse than those who focused on strategy—even when under pressure. All of the soccer players with whom Jackson worked were of equal caliber. These findings show what Jackson calls the paradox of control: athletes focus on elements of their technique that they believe will help enhance performance. Paradoxically, this technique focus results in worse performance than if they paid no attention to detail in the first place. Getting athletes to focus instead on performance strategies helps to take their minds off what they are doing and ensures they play well.

Some psychologists have suggested that focusing on a key word related to the outcome of the intended play or action is the best medicine for poor performance under stress. Top golfers were asked to putt while thinking about a word that encapsulated the entire putting motion (such as *smooth*) instead of words that represented physical aspects of their technique (such as *head, knees,* and *arms*). When just putting in a practice situation where there was no pressure to succeed, the golfers with holistic swing thoughts performed the best. More importantly, when the pressure was ratcheted up by offering cash to the best putters, the performance of the golfers who focused on the holistic key word was not hurt while the golfers who focused on technique choked under the pressure.[12]

A one-word mantra can keep golfers focused on the end result rather than the step-by-step processes of performance. When this mantra relates to performance outcomes, it works even better, because when you perform skills your have mastered, it's almost as if your brain knows what to do. Focusing on the outcome of your actions—using words like *smooth* or *net*—helps the brain organize the processes you need in order to actually produce the end result. Think of focusing on an outcome as working backward. How many times has a college basketball player done the same flicking movement of her wrist to sink her favorite shot from the right side of the key? Hundreds of thousands of times. Because of this experience, when our basketball player sees the net from this angle, the visual image of the net actually cues the motor areas in her brain to put the correct wrist actions together to sink her signature shot. The mirror network is engaged.

Sports scientist Gabriele Wulf talks about focusing on the outcome as an external focus of attention. She suggests that, rather than paying attention to your body (an internal focus of attention), athletes from surfers to skiers to skaters to soccer players should pay attention to some aspect of their environment—where they want to go rather than where they are right now. Phrases like "focus two gates ahead" in skiing or "focus on the empty net" in soccer epitomize this view. Focusing on the outcome of actions may be all a player needs to do in order for the well-trained brain and body to produce it.[13]

THOSE WHO CAN DO SHOULDN'T TEACH

Yogi Berra is not only famous for his ability to hit baseballs out of the park, he is also known for his "Yogi-isms"—amusing quotes Berra has made over the years that consistently make half the people who hear them laugh and the other half scratch their heads in perplexity. One of my favorite Yogi-isms is about hitting: "How can you hit and think at the same time?"

Although we may never know what Yogi really meant, I have always thought that he was trying to get across the idea that highly skilled athletes perform at the top of their game when they are not overanalyzing their performance. Of course, this idea is not limited to Yogi Berra. Hall of Fame Chicago Bears football player Walter Payton has remarked that when he was on the field he didn't know what he was doing. "People ask me about this move or that move, but I don't know why I did something. I just did it."

Although beginners must attend to their performance in order to ensure they don't make silly mistakes, once they've learned a skill well they don't continue to pay the same attention to detail. As a result, asking skilled athletes to comment on what they just did can be bad for the performers themselves and for anyone to whom they are trying to teach their skill. Oftentimes, the best players make the worst coaches.

Take a recent study conducted by psychologists Mike Anderson and Kristin Flegal.[14] The researchers asked highly skilled and beginning golfers to take some short putts on a fairly flat, straight green. The golfers then spent several minutes describing the putts they had just taken, or they worked on an unrelated task instead. Afterward, all the golfers were asked to perform the putts again. After spending time describing their past putts, the skilled golfers needed twice as many attempts to sink their putts as the skilled golfers who had not put their performances into words. Beginning golfers' performances were not affected by describing putts. These less skilled players even improved a little bit when asked to recount what they had just done.

For well-learned activities like taking a free throw, hitting a simple putt, or playing a cadence that you have performed a thousand times in the past, thinking too much about the step-by-step processes of what you are doing can be detrimental. Not only can trying to describe performance disrupt it, but skilled athletes often have trouble putting their actions into words in the first place. When golfers are pushed to describe their recent actions, they can't explain them. For instance, when a scratch golfer in my lab was asked to describe a putt he just took, he replied, "I don't know. I don't think while I putt." When your performance flows largely outside your conscious awareness, your memories of what you've done are just not that good.

Maybe this is why the best players are usually not the best coaches. According to Canadian hockey player Thérèse Brisson, a member of the gold-medal-winning women's team at the 2002 Winter Olympics in Salt Lake City, "Recently retired hockey players who played at high levels rarely make the ideal coaches for youth hockey. They know what to do, but they can't communicate how they do it!" She says that given the choice between a skilled hockey player and an experienced physical education teacher to help at the youth hockey camps she now runs, she will always take the teacher. "Teaching skating skills is one of those problem areas," Brisson says. "How exactly do you skate faster?" Being able to communicate this type of information comes from coaching experience, not from playing experience.[15]

CHANGE WHAT YOU FEAR

To be honest, even if you use all the psychological techniques at your disposal to deal with pressure's negative effects, inoculation against the choke is never a sure thing. Unfortunately, once athletes start to crack under stress, it can be hard to stop doing so.

For instance, think about a runner who stumbled during the five-hundred-meter dash at the U.S. Olympic trials. Just because her performance is over, her poor showing will not be forgotten—especially by the runner herself. As anyone who has ever choked knows, these types of flubs can haunt you. Sometimes that one choke can even end a career, but many athletes can turn a poor performance around and eventually start playing well again.

What determines whether choking will become a problem or a one-time phenomenon never to be repeated again? This is a question Olympic gold medalist Lanny Bassham faced. A marksman and admitted choker, he says he caved under the pressure in the 1972 Olympics. But rather than giving up, Bassham began talking to every Olympic champion he could find in the hope of figuring out what it takes to win. Bassham's hard work paid off and he came back to win a shooting gold at the 1976 Olympics. Since then, Bassham has been working with PGA tour players and other athletes, taking what he learned about success and failure as a shooter and extending it to the links, the court, and beyond.[16]

One major point Bassham tries to pass on to the guys he works with is that focusing on the negative or on what you might lose if you don't succeed is one of the worst things an athlete can do. Excessive negative self-talk really hurts your performance. And it won't hurt just your immediate performance. Negativity can also impair performances down the line. In working with athletes, sports psychologists often emphasize the importance of losing any negativity.

Hap Davis, the team psychologist for the Canadian national swim team, has even designed a type of intervention that helps swimmers think about their failed performance in a more positive light and improves subsequent swims by changing how these athletes' brains deal with the negativity. To do this, Davis teamed up with a group of neuroscientists in Canada to peer inside the heads of Canadian national swimmers while they thought about swims during which they had choked.[17] Davis's goal was to see if he could turn the brain's reaction to a failed performance into something positive. In one study, Davis and the researchers used fMRI to look at about a dozen swimmers who'd failed to make the 2004 Canadian Olympic team at the pressure-filled Olympic trials. They also looked at some other swimmers who made Team Canada but performed poorly under the pressures of the Olympic stage.

The athletes watched videos of their failed races and videos of other swimmers competing. While they watched, their brains were scanned so that the researchers could see the type of brain activity elicited by the swimmer's own failed performance. Following the videos, the athletes received a short intervention designed to help them reframe their thinking about their bad swim and put it in a new light. Then the swimmers watched their failed performance again.

The intervention had three parts. The swimmers were asked to 1) express the feelings they had when they watched the failed race; 2) think about what went wrong in the actual swim (for example, "I was slow," "I need to work on my stroke length"); and 3) imagine performance changes for the next race.

When the swimmers initially watched their own poor performances before the intervention, they showed more activity in several emotional centers in the brain, such as the amygdala, than when they watched videos of other competitors racing. They also showed increased activity in the prefrontal cortex, which, as we learned, seems to be a major culprit in paralysis by analysis. Finally, the researchers found decreased activity in motor regions of the brain that are essential for the planning and execution of movements. Davis and his research team think that the decrease in motor activity in response to the swimmers seeing themselves fail may be similar to what is observed in animals when they are trying to escape and know there is no way out. In these situations, the animal stops trying and simply lies still. This is sometimes referred to as learned helplessness, a phenomenon in which people don't feel they have control over a particular situation or outcome so they stop working to try to obtain a goal.

But when the swimmers watched their failed races again after the intervention, they had less emotion-related brain activity and more activity in important motor regions of the brain. By getting rid of the negative emotions stirred up by reliving a defeat, the short intervention may have helped to prime the brain for action and improved competition.

It's not uncommon for a swimmer to blow a race early in a meet and then perform poorly in other events in the rest of the meet because they are dwelling on their first failure. This occurs in other sports, too. Gymnast Alicia Sacramone might have benefited from practices of the Canadian swimmers. Perhaps if she had been able to quickly assess what went wrong on the beam, imagine herself performing her upcoming floor exercise, and move on to highlight something positive about her next event this could have pushed her to excel and have given the U.S. the all-around women's gymnastics gold.

These sorts of techniques seem to be working for Team Canada, whose sport psychologist Davis has a new policy of immediately picking up on any negativity. Davis has swimmers immediately review a bad performance poolside and think about how they would fix it. As a result, Canadian swimmers are turning around their performances right away.

MEDICATING THE CHOKE

In the 2008 Beijing Summer Olympics, there were the anticipated Olympic moments of triumph, glory, and of course, defeat. There were also the typical instances of athletes being thrown out of the games for doping. Some even lost medals because of it. Ukrainian heptathlete Lyudmila Blonska was stripped of her silver medal, North Korea's Kim Jong Su was stripped of both silver and bronze medals in pistol shooting, and the International Olympic Committee banned U.S. track and field star Marion Jones because of her past doping issues from even competing in the games.

Although many athletes were thrown out for using anabolic steroids like tetrahydrogestrinone (THG)—now a household name in the United States thanks to the baseball doping scandals of the last few years—not all the doping cases involved steroids. Equestrian riders from Norway, Brazil, Ireland, and Germany were thrown out because their horses tested positive for the banned substance capsaicin, which provides the heat in chili peppers and is prohibited because it can be used as a pain reliever. Because a horse can't tell his rider if something hurts, even a mild pain reliever can allow the horse to be pushed too far and lead to serious injury. As a result, capsaicin and its relatives such as Nonivamide are banned in Olympic competition.

North Korean pistol shooter Kim Jong Su was thrown out, not for steroid use, but for using a drug called propranolol. Propranolol is in a class of drugs known as beta-blockers, which lower blood pressure by blocking particular sympathetic nervous system receptors, such as the receptors for adrenaline, often talked about as the "fight-or-flight" hormone that is released in times of acute stress.

Drugs that block the binding of adrenaline with adrenergic receptors may be useful in some sport and performance endeavors, because when people take beta-blockers, their hands shake less, their voice doesn't tremble as much, and they are less likely to exhibit outward signs of anxiety. Beta-blockers have been banned in such sports as archery and pistol shooting because of the advantage they may confer in stressful competitions.

A study conducted in the mid-1980s suggests that beta-blockers do work.[18] A group of Dutch scientists asked about thirty skilled marksmen to shoot a standard 25 mm pistol after taking either the beta-blocker metoprolol or a placebo drug. The study was double blind—neither the researchers nor the shooters themselves knew who had been given the drug and who had been given the placebo. Shooting was about 13 percent better in the marksmen given metoprolol compared to those who got the sugar pill. Interestingly, there was no relation between shooting improvement and heart rate or blood pressure changes. Rather, metoprolol seemed to improve shooting by alleviating hand tremor.

Beta-blockers are used by dancers, musicians, and archers, as well as by people who are afraid of public speaking. The role of beta-blockers is not completely understood. Rather than improving overall performance per se, they may allow those who get most anxious in competitive situations to show what they can do. If so, maybe beta-blockers are not so bad.

In work with skilled musicians, researchers gave the beta-blocker oxprenolol to string players from several London colleges.[19] They then upped the pressure by asking the musicians to perform while they were scored by professional judges. Overall, the musicians trembled less on the beta-blocker and received higher performance scores. But a closer look reveals that oxprenolol didn't just improve performance across the board. Rather, it was those players most prone to stage fright—the anxiety and stress that can come from performing in public—who improved with the help of beta-blockers, most likely because the drugs reduced tremor. Players who did not normally get stage fright didn't benefit from oxprenolol at all.

Professional musicians, like athletes, have long recognized that excessive anxiety can impair performance, sometimes with catastrophic results. Yet while beta-blockers are banned in Olympic sports like pistol shooting, they are accepted in the musical world. Some argue that beta-blockers help nervous performers play up to their potential, but others believe that performance-enhancing substances of any kind have no place in sports or on the stage. Of course, those whose fear of performing would prevent them from otherwise playing do not want to include beta-blockers with the rest of the banned, performance-enhancing drugs.

WHERE ARE WE NOW?

Coaches often admonish their athletes to "get their heads in the game." Indeed, after David let the second penalty shot slip right through his hands at the Illinois boys state championships, this is exactly what his frustrated coach had yelled from the sideline. But this advice may not always be a good thing. As we have touched on throughout the last few chapters, overlearned actions operate largely outside working-memory. And when we attempt to bring these actions back under conscious control, performance often suffers.

Whenever players find themselves needing to perform skills that they have executed flawlessly in the past, their goal should be to become less conscious rather than more. They need to get into the *flow*, a concept proposed by psychologist Mihaly Csikszentmihalyi to describe the mental state of being fully immersed in what you are doing—as if you were in a current that was carrying along. The term being *in the zone*, as sport psychiatrist Michael Lardon calls it, also describes the beneficial loss of the feeling of self-consciousness.

At age fourteen, golfer Kimberly Kim became the youngest champion of the U.S. Women's Amateur Golf Tournament. And she did it with a bang. On the last tournament day, she had three birdies to finish the morning round. When asked to describe how she did it, she replied, "I don't know how I did it. I just hit the ball and it went good."

Playing "out of one's mind," so to speak, is likely one of the reasons that professional athletes don't often give the most informative interviews after their big game. They can't tell you what they did because they don't know themselves and end up thanking God or their mothers instead. Because these athletes operate at their best when they are not thinking about every step of performance, they find it difficult to get back inside their own heads to reflect on what they just did.

A number of techniques (highlighted on the next page) help performers play at their best and deal with the pressures they often face during the solo performance or the game-winning kick, free throw, or shot—techniques such as speeding up the execution of a well-learned skill, focusing on the outcome rather than the process, and even distracting yourself. A method on which most people agree is that practicing under the types of pressure you will face in actual game situations is one of the best antidotes for the negative effects of stress. You will see that this is a truism in the next chapter as well, where we talk about performance under pressure in the business world.

Tips to Combat Performance Flops Under Pressure in Sports and Performance

Distract yourself. Singing a song or even thinking about your pinky toe as Jack Nicklaus was rumored to do can help prevent the prefrontal cortex from regulating too closely movements that should run outside awareness.

Don't slow down. Don't give yourself too much time to think and to control your highly practiced putt, free throw, or penalty kick. Just do it.

Practice under stress. Practicing under the exact conditions you will face in a do-or-die situation is exactly what is needed to perform your best when the stress is on. Get used to the pressure so competition is not something you fear. Also, by understanding when pressure happens, you can create situations that will maximize the stress in your opponents.

Don't dwell. Take that past performance and change how you think about it. See your failures as a chance to learn how to perform better in the future.

Focus on the outcome, not the mechanics. Focusing on the goal, where the ball will land in the net, helps cue your practiced motor programs to run flawlessly.

Find a key word. A one-word mantra (such as *smooth* during a golf stroke) can keep you focused on the end result rather than the step-by-step processes of performance.

Focus on the positive. Don't be helpless. If you focus on the negative this can make you feel out of control and increase the likelihood that you will not work as hard to obtain future performance goals.

Cure the yips by changing up your grip. An alteration in your performance technique reprograms the circuits you need to execute your shot, hopefully clearing your brain and body of the motor hiccup.

CHAPTER NINE

CHOKING IN THE BUSINESS WORLD

Ed was not what you would call "interview savvy." Although he came off as thoughtful and put-together over e-mails when he had ample time to ponder what he was going to say, if he had to give a phone interview or go in for a one-on-one meeting, everything seemed to fall apart. Even when Ed was asked a question that he knew how to answer, he would often stumble on his words or his mind would go blank completely. This never went off well with interviewers, especially since Ed was in public relations—a job where polished communication skills are essential. Despite his credentials, a Harvard undergraduate degree and a business degree from the University of Chicago, Ed had lost more than a few jobs because of his lackluster interview performance.

The nerves usually started the night before the meeting. Ed wouldn't sleep well and he would wake up too tired and nauseated to eat breakfast. By the time Ed actually got to the interview, he was already worn-out. There was no way he would be able to think at his best. Ed would inevitably forget something he was supposed to know about the company or answer a question in a way that didn't go over well with his interviewers and, although recovery was possible, Ed would feel like the meeting was ruined. This preoccupation with his mistakes was like a snowball effect. Ed would be distracted, miss more questions, and make things worse.

In this last chapter we step from the classroom, playing field, and performance stage into the business world and look at several other potentially stress-filled situations—from interviewing for a job where you have one shot to impress, to giving talks or sales pitches when all eyes are on you. We will talk about how these types of high-stakes activities are both similar to and different from the sport, academic, and performance situations we have discussed thus far. The goal is to understand why people choke in activities common to the business world so that you can find the right technique to alleviate poor performance under stress in your own pursuits.

FIRST IMPRESSIONS

The other night I watched a movie called *Swimming Pool,* a 2003 thriller that takes place in a villa in southern France. Middle-age British crime writer Sarah Morton (played by Charlotte Rampling) has run off to the villa, owned by her publisher, to find solitude and inspiration to write her next book. All is well until—unexpectedly—the publisher's wild teen daughter Julie (played by Ludivine Sagnier), arrives to stay. Julie has a different man in her bed every night and, not surprisingly, clashes with the uptight Morton. The entire affair ends with a murder and a cover-up. Importantly, at the very end of the movie, there is a plot twist in which you, the viewer, learn that the publisher's daughter is a different person from the girl who showed up at the villa in France. The viewer is left to assume that the girl in France was a figment of the writer Sarah Morton's imagination, concocted in her head so that she could write a book about the events that unfolded at the villa.

A friend of mine had picked the movie and, although he had seen it once before, he professed to remembering very little and was happy to watch it again. Interestingly, my friend had completely forgotten about the plot twist introduced at the end of the movie, which changes how you see the entire film. How, he wondered, could his memory have lapsed concerning this major movie detail? Some classic psychology research can help explain this phenomenon and will likely help people like Ed ace his interview as well.

Our ability to make sense of movie plots, to navigate novel situations, or even to form first impressions of the people we meet is greatly aided by what we psychologists call *schemas*. Packets of knowledge that provide expectations about the activities we do, schemas help us comprehend new situations with familiar details. For instance, we all have a schema for what happens in a restaurant. We expect that when we go in to, say, a new pizza place, we will be seated, a waiter will take our order, someone will bring us a pie, and we will be expected to pay for this service before we leave. If we didn't have this restaurant schema, upon entering a dining establishment we might proceed directly to the kitchen and start cooking ourselves. Schemas help us make sense of new situations we encounter based on what we have learned about similar activities in the past.

Schemas help us interpret new activities or situations in a meaningful way, but this only occurs if we are given the schema before (not after) we encounter the novel information. The reason my friend had completely forgotten about the movie plot twist is that it came at the *end* of the movie. If the plot twist had come about earlier, this would have provided him with a schema for interpreting the flick (as a movie that is a figment of the main character's imagination) and he would have remembered things very differently.

A classic psychology experiment conducted in the 1970s illustrates the power of getting a schema or interpretive guide before you encounter a novel situation. The results of this experiment can help you ace your interview as well. Students were asked to read the passage below and then, after having read it, recall as many details as possible.[1]

> The procedure is actually quite simple. First, you arrange things into different groups. Of course, one pile may be sufficient depending on how much there is to do. If you have to go somewhere else due to lack of facil-

> ities, that is the next step, otherwise you are pretty well set. It is important not to overdo things. That is, it is better to do too few things at one time than too many. In the short run, this may not seem important but complications can easily arise. A mistake can be expensive as well. At first, the whole procedure will seem complicated. Soon, however, it will become just another facet of life. It is difficult to foresee any end to the necessity for this task in the immediate future, but then one can never tell. After the procedure is completed one arranges the materials into different groups again. Then they can be put into their appropriate places. Eventually, they will be used once more and the whole cycle will then have to be repeated. However, that is a part of life.

Perplexed? When people were told nothing about the passage ahead of time, they were confused about what they were reading and their memories for the details of the passage were pretty poor. However, when people were told ahead of time that the passage was about *washing clothes,* their memories for the passage substantially improved. Interestingly, if people were told that the passage was about washing clothes after they read the passage, but before they were asked to recall the details, their memories were no better than those of individuals who had never been told what the passage was about in the first place. The take-home point is that having the appropriate schema or context for encoding information helps us understand and recall this information, but only if we get the schema at the outset.

Schemas are relevant to interview situations because giving your interviewer a positive schema for interpreting your employment potential early on in the meeting can help shape how he or she remembers the entire encounter. If you start out with a few well-rehearsed sentences about why you are the right person for the job, this first impression can help set the tone for your interview and for what is taken away from the meeting.

We can look to the brain to understand why first impressions might carry so much weight. When we initially have an experience, whether it is watching a new movie or meeting someone for the first time, we recruit a network of brain regions to help make sense of what we encounter and to help us store this information away in memory. Schemas determine how this new information is stored and what is actually remembered.

Comparing brain activity when people actually remember pictures or words they saw previously versus when they do not, scientists have been able to get a pretty good idea of what brain areas are responsible for accurate memory and when these memory traces get laid down. For instance, brain activity in the medial temporal lobe and prefrontal cortex when people first encounter a new situation predicts subsequent memory accuracy.[2] As we have discussed in past chapters, the prefrontal cortex houses working-memory, which is instrumental in guiding what we attend to and what we ignore. The prefrontal cortex uses schemas or prior expectations as a guide for what to focus on and what to take away from an initial meeting. Providing a schema for interpreting a meeting at the outset, then, can help guide others' memory of you.

First impressions are important. Set the stage early on for what your interviewers remember about you by giving them a positive schema by which to encode your job potential. Even if you show nerves after the fact, this initial impression may help ensure your success.

MIMICRY

How else can you make a good impression in an interview? A few chapters back we discussed how it is that expert athletes are able to anticipate the actions of others. Because experts run off—in their own brains—a simulation of what others are doing, these athletes can start to react before an action by their opponent has been completed. The concept of mirror system was introduced here; the idea that part of the same neural circuits involved in doing are also used for perceiving and understanding. If, when athletes see someone else performing an action, they themselves call upon their own motor repertoires to play out this action in their own head, these athletes will be privy to information about how the action is likely to turn out—even before the action is finished.

As it happens, one way to engender positive feelings about you in others is to act as they act, and this too is related to the mirror system. The idea is simple: if you and your interviewer are both making the same movements, you will be better able to interact smoothly with them because your motor systems are in sync. When your interviewer has crossed arms and you cross your arms, he or she is better able to make sense of what you are doing because he or she is able to "mirror" your actions on his or her own motor repertoire.[3] And when we feel like we are really in tune with someone else, we like him or her better.

Psychologists Tanya Chartrand and John Bargh were at New York University when they first discovered this mimicry phenomenon.[4] They called it the *chameleon effect*—we often unconsciously mimic the postures, mannerisms, facial expressions, and other behaviors of our interaction partners, and this engenders liking.

To show the chameleon effect in action, Bargh and Chartrand invited unknowing NYU students to their laboratory and had them participate in an activity similar to an interview in which two people interacted one on one. People took turns describing various color photographs taken out of *Time, Newsweek*, and *Life* magazines. The students were told that the researchers were interested in creating stimuli for a test that, similar to the Rorschach test, could be used to gain insight into the personality and motivations of clinical populations. The college students had purportedly been recruited to describe what they saw in the pictures before the scientists asked people with particular pathologies (depression, mania, etc.) to project their thoughts onto the same photographs.

In actuality, however, the photo description task was just a cover story. Unbeknownst to the students, one of the two people asked to describe the pictures was actually a confederate, a researcher working in the lab who had been told to make particular bodily movements during the one-on-one interaction. The movements that the confederate made included things like shaking his or her foot or rubbing his or her face.

The researchers found a chameleon effect. When confederates rubbed their faces, so did the student, and when confederates shook their feet, the participant did the same thing. This was true of facial expressions as well. The NYU students smiled, on average, a little over once a minute when they were with a smiling confederate and averaged only a third of a smile per minute when they were with a confederate who did not smile. We judge people and objects to be more pleasant when we are smiling in comparison to when we are frowning, so if you want your interviewer to think positively about you, try smiling. The saying is true: "when you're smiling, the whole world smiles with you."[5]

The NYU researchers turned the tables in a second study. Again they asked students to interact with another person who was really a confederate, but this time, the confederate either mirrored the mannerisms of the student or did not. For instance, when the student crossed his or her legs, so did the confederate. Interestingly, students reported smoother interactions with the confederate when the confederate mimicked their behaviors versus when they did not and the student reported liking this ostensible stranger more. The NYU students were completely clueless that their positive feelings toward this person had anything to do with the similarity of the behaviors, but they did. Students liked the confederate more when their mannerisms were in sync.

In interview situations, regardless of how nervous you are, your behavior counts. Keeping positive and being a chameleon with your interviewer can help. Indeed, other research has shown that when interviewees mimic the gestures and mannerisms of interviewers, the interviewer believes the interviewee is better informed and has sounder ideas than when he or she does no mimicry. Of course, don't take this too far. Once people are aware that they are being copied, the liking can turn into annoyance.

Mimicry may also be the foundation for good interpersonal relationships. Mimicking the facial expressions of your partner is good for your marriage, because when you imitate others' emotional expressions, your brain is in a better position to understand what emotional state they are in and, as anyone who has been in a long-term relationship can tell you, the ability to empathize with your partner is key.

Couples often start to look alike as they grow older, because one consequence of mimicking a partner's facial expressions after years of cohabitation is that the repeated use of the same facial muscles means that faces start to look more similar. If one partner smiles in a certain way and the other is likely to copy it, similar patterns of wrinkles and molding of facial muscles will occur. Because empathizing is likely one key to marital bliss, it follows that spouses who look more alike after many years of marriage—because of the mimicking of expressions and mannerisms—should report being happier together. This is exactly what research has found.

In one study, more than one hundred people were shown photographs of men and women in their first year of marriage and then again of the same couples twenty-five years later—on the spouses' silver wedding anniversary.[6] The researchers went to a lot of effort to remove extraneous information and crop the photos so that only the couples' faces could be seen. People were asked to judge the physical similarity of the couples.

The researchers found an increase in similarity of appearance over a quarter of a century of marriage. This finding can't just be explained by the idea that all people generally look more alike as they get older, because when the researchers randomly matched older couples together, they weren't rated as any more similar than randomly matched younger couples.

To see if the couples who looked most alike at their silver wedding anniversary were really the happiest, the researchers mailed surveys to the couples and asked them to rate their marital satisfaction. Each questionnaire was mailed separately to the husbands and wives to minimize the chance of one spouse influencing the other's responses. The greater the resemblance increase over the twenty-five years of marriage, the higher the couple's reported happiness. Thus, being in sync with your spouse or your interview partner can help ensure that your emotions and viewpoints are on the same page. The ultimate result is greater liking, longer relationships, and, in the business world, an increased probability of you landing a job.

THINK ABOUT WHAT TO SAY, NOT WHAT TO AVOID SAYING

When Sheila went for an interview as a midlevel manager at an advertising company, she was not visibly nervous. But she had to admit that she didn't feel as if she had shined when asked tough on-the-spot questions. One reason was that she was afraid she would blurt out something she shouldn't say: a comment about how the interviewer was dressed or inside information she had learned about the company. Sheila tried to take her mind off these things, but she always felt like she was half distracted by trying to *not* think. Sheila was doing exactly what she shouldn't do. When you have a glass full of red wine and find yourself at a dinner party walking across the hostess's new white carpet, making a note to yourself "not to spill" may actually bring about the unwanted occurrence. Trying not to think about something may result in the propensity to have the exact thought you are trying to avoid.

Daniel Wegner, a psychologist at Harvard University, has spent much of his career researching how it comes to pass that the very thoughts and actions people try to avoid end up happening. Wegner suggests that there are really two processes at work when we try not to think about something. There is a conscious process that searches for some new topic to focus on. Then there is also an unconscious search for the unwanted thought, whose purpose is to check for errors in our ability to strike the unwanted thought from mind.[7] Together, these two processes help people avoid topics they don't want to focus on and, most of the time, people do this pretty well. But the story changes when we find ourselves under the gun. Pressure to perform well often attacks the prefrontal cortex—the very seed of the conscious processes that search for new topics to think about. Under stress, then, Wegner argues, we have only the unconscious working for us, the process actually charged with finding the very thing we don't want to focus on. As a result, we are likely to blurt out exactly what we are trying not to say or make the move we tried to avoid. Trying not to think about something in a stressful interview may be the worst thing you can do.

In sports, some have argued that the yips are particularly likely to occur when athletes are trying hardest to avoid them. Similarly, my research team and I have shown that when golfers are told not to putt the ball long (or short), they are more likely to mess up.[8] Soccer players instructed to avoid kicking a penalty shot within reach of the goalkeeper are more likely to focus their gaze on the goalkeeper and shoot right at him.[9] Just think about baseball player Chuck Knoblauch and his famous wild throws to first base. It was well-known that Knoblauch had an obsessive desire to avoid these throws and this may have ironically increased their occurrence.

What should you do if you find yourself thinking about exactly what you are trying to avoid? There are several ways to combat these ironic effects—some of which we have already talked about as tools for combating stress more generally. For instance, meditation may help you learn not to dwell on what you want to avoid, especially when under pressure. Meditation may be helpful in that people learn to recognize the unwanted thoughts and let them go, rather than suppress them. Letting thoughts go means a lower likelihood of pop-ups in the future. In addition, writing down thoughts you want to avoid (just as writing about your worries before a big test) may help to disclose these thoughts and make them less likely to occur when they are least wanted. Finally, when soccer players are instructed to focus on a particular spot in the goal and kick the ball there, they are good at doing this. People look where they aim and they aim where they look. On the pitch, direct yourself to focus on the empty net rather than on the goalkeeper, and in the boardroom, focus on your three strong talking points rather than what you don't want to say.

SPEAKING IN FRONT OF OTHERS

Public speaking is so much a part of every businessperson's life, whether in a boardroom or business conference, in an open forum or just interacting as a team leader. This is why Linda decided she had to do something about the pressure she felt when she found herself in front of an audience with all eyes on her. A member of a New York–based real estate investors group, Linda became nervous when pitching to potential clients but she also stressed out in more informal situations where she was given only a few seconds to get her thoughts out—say, on an elevator ride down to the ground floor with her boss.

Giving a presentation to a roomful of people or just voicing your opinion in an elevator to someone high up in your company can be a stressful experience. Even if we are prepared, have made notes, or practiced our introduction, once we are on the spot it's pretty easy to get unnerved. Why does it happen, even if it is just once in a while? So much has been written about how to give an effective presentation. Sometimes folks who call themselves public-speaking experts tell us to think about every word; others say to try to make our minds blank. What is the right strategy and why do some people seem to have no fear while others can barely make it to the stage? Let's get some answers to these questions by taking a look at what happens when people prepare to talk in front of others.

For over fifteen years, researchers around the world have been inviting people into their laboratories for the sadistic purpose of stressing them out by asking them to prepare a speech that they will have to give in front of others. The test is called the Trier Social Stress Test, named after the university, Trier University in Germany, where it was developed.[10]

The test goes something like this:

Upon arrival at a research lab, participants are led into a room occupied by a three-member panel. People are asked to take a seat opposite the panel and told that they have the task of creating a five-minute presentation that will convince the panel that they are the best applicant for an open job in the laboratory. People are told that they will be evaluated on both the content of the speech they create and their presentation style—this means no fidgeting, "ums," etc. The person usually has about ten minutes to prepare the speech. Then, with a video camera focused on them and their every move, the study participant is asked to actually stand up and give the speech to the oftentimes less than supportive panel situated in front of them. As if that were not enough, immediately after the speech the person is asked to do another task where he counts backwards from 1,022 by 13—out loud—as quickly and accurately as possible.

Taking part in the Trier Social Stress Test can be nerve-racking. And a decade and a half of research has shown that, for most people, this public speaking activity is a clear, reliable way to elicit a substantial cortisol response, the hormone that is a marker of stress. Yet it's not just giving a speech or doing math that induces the most stress; the Trier Social Stress Test must include elements of social evaluation (a panel judging you) to strongly and reliably cause a reaction. Judgmentalism is a mainstay of most public-speaking conditions; people are afraid of being evaluated and looking like a fool.

Despite the pressure-filled nature of the Trier Social Stress Test, however, research has also shown factors that can help take the pressure off public speaking. In one recent study conducted on over one hundred people at the Centre for Studies on Human Stress in Montreal, researchers found that people with more education, a university degree for instance, were less affected by the Trier Social Stress Test than those who did not have a postsecondary school experience.[11] Because going to college usually means exposure to myriad stressful situations—from having to speak up in discussion sections to working out problems in a study group—less education translates into a lower likelihood of having experienced stressful public-speaking situations in the past. Getting used to the pressure of speaking in front of others helps people not react as negatively in the future. Becoming accustomed to the pressure of performing in front of others makes public speaking a lot less daunting.

It's well-known that when President Obama has to give a big speech, he practices extensively ahead of time. When it's actually time to step up to the microphone, he knows what to do. Even though a majority of the free world is hanging on Obama's every word, he is cool under pressure precisely because he has practiced performing in front of a crowd. Obama knows how to sink the game-winning shot at the podium and also, as Hyde Park lore goes, on the court as well. He does it with practice.

It is not necessary to practice with the specific speech or pitch you have to give, but practice in general can help. For instance, if you spent time each week making a fool out of yourself—perhaps by taking an acting class or doing improvisation or just giving toasts with your friends—this experience might help you alleviate your fear of speaking out loud. When you know what the worst thing that can happen is, and you have experienced it already, you will be less likely to worry about it.

People who have strong social support systems are also less likely to be stressed out by the prospect of giving a speech. Researchers at the University of Zurich have found that men who were either married or living with a significant other and able to spend time with their spouses before having to prepare their speech showed less of a cortisol increase in anticipation of the Trier Social Stress Test than those who didn't spend time with their spouse. There are a few caveats, however, to these results. Having a spouse present was only beneficial for the speaker's anxiety levels if the speaker was in a healthy relationship. The opportunity to spend time with your spouse before a stressful performance situation, when you are not in a good place with this person to begin with, is actually more destructive than if you were alone.[12] Moreover, research conducted with women and their partners shows that the social support benefits that men experience don't always hold for the opposite sex. Indeed, in one study, women's cortisol levels went up during the Trier Social Stress Test when their boyfriends were present beforehand relative to when he was not.[13] Whether this was due to a lack of support on the part of the boyfriend or an inability of the women to receive it is still open to debate.

Having chronic stressors in your life—whether it's a rocky marriage, out-of-control kids, financial worries, or aging parents—can send you over the edge when you have to speak in front of others. That's why having your partner with you before a speech is beneficial if and only if the partner is not one of your major pressure inducers. The classic inverted-U function explains why the presence of a partner can either be helpful for reducing stress or can backfire completely.

The inverted-U depicts a common relationship between performance and arousal.[14] As you can see, being aroused up to a certain point is good for performance because it gets you motivated and energized to succeed. But once you are at the top of the U, increased pressure is not a good thing since it will send you crashing over the edge. People with chronic stress in their lives are likely to sit at the top of the U under normal conditions, so when they are faced with the added pressure of public speaking they may be more likely to perform poorly than those who normally sit on the uphill side. If a spouse who is anything but a calming entity is put into the mix, the consequences can be disastrous.

Whether you are pitching ideas to other VPs or leading a meeting with potential clients, chronic stressors outside your work life can creep in. Just as putting pen to paper helps to curtail the power of worries before a big test, getting chronic stressors out on paper can be beneficial as well. Merely writing about the stressful events in your life on a regular basis—say, twenty minutes once every week or so—can

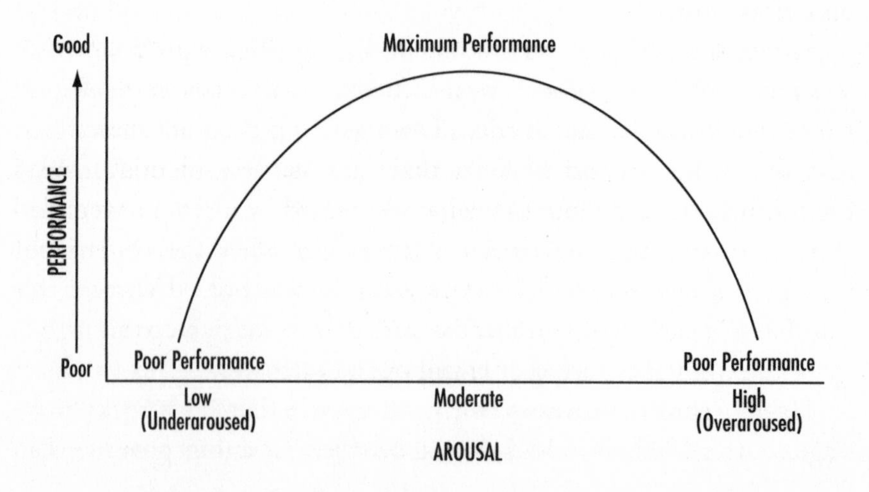

bolster your cognitive horsepower by decreasing the occurrence of intrusive thoughts and worries.[15]

Think of a computer analogy: if a computer is running several programs at once, each one of these programs will run that much slower and be more prone to crash. Getting your worries out on paper eliminates the unnecessary worry-programs from running and helps you focus on the task at hand.

Following are some of my favorite recent speech and interview situations where politicians' performances went awry and where some of the techniques we have talked about for dealing with the pressure may have helped save the day.

No one can forget Republican vice presidential candidate Sarah Palin being interviewed by CBS News's Katie Couric. During this exclusive interview, Couric pushed Palin about the financial crisis in the United States and asked her whether the lobbying firm of Senator John McCain's campaign manager, Mark Buse, had received money from mortgage giant Freddie Mac. Palin fumbled over her answers and it only got worse from there. It was rumored that Palin refused to be worked on by McCain's "handlers" and thus had not practiced answering questions under the gun. The vice presidential candidate needed all the working-memory she could muster for her interview and, because she hadn't prepared ahead of time for on-the-spot questions, her mental activity seemed to go out the window when she needed it most.

Howard Dean's scream: Talk about letting go of prefrontal cortex control. After losing the Iowa Democratic caucuses in the run-up to the 2004 presidential election, Dean gave a passionate speech that involved, at the very end, a scream that was not so presidential. Instead the scream sounded more like what you would hear from a wounded animal backed into the corner. Played over and over on twenty-four-hour news channels for days afterward, the howl was a likely cause of the early front-runner's demise. Even when we are fatigued and in a high-stakes situation, engaging the prefrontal cortex to inhibit unwanted behaviors is important.

Take the 1992 town hall debate between Arkansas governor Bill

Clinton, Ross Perot, and Bush senior. When you tell yourself not to do something, ironically the action can be more likely to occur. President George H. W. Bush has a habit of checking his watch in important situations and no doubt his handlers had told him to try not to do this action on the stage. In the middle of the debate, however, all cameras and eyes were on Bush as he glanced at the time. Checking your watch is perhaps not the best move if you want people to think that their president is interested in what is going on and not anxious to flee the scene.

In the 1988 presidential debate between Vice President George H. W. Bush and Massachusetts governor Michael Dukakis, Dukakis was asked a question about his anti-capital-punishment views, specifically with respect to a hypothetical scenario in which his wife was murdered and raped. Dukakis flubbed the response, coming off cold, flat, and negative in a situation where he had a real chance to show a more sensitive side. To be fair, Dukakis had the flu and was likely not functioning at his best, but his inability when he was pushed into a corner to present a compassionate schema for himself may have cost him the election.

ANTICIPATION

Merely preparing to give a speech that will be evaluated by others can be enough to send most people's anxiety skyrocketing, as psychologist Tor Wager and his colleagues at Columbia University have recently shown.[16] Wager was interested in what happened in the brain in the run-up to an important and pressure-filled public-speaking situation so he put Columbia students in an fMRI machine and informed them that they would be given a few minutes to mentally prepare two different speeches—one on the effects of interest rates on stock prices and the other on the relationship between tariffs and free trade. The students were told that they would be presenting the speeches to a panel of experts in law and business and that a computer analysis program used to grade college-level essays would also score what they said. In reality, and to the students' relief, they never had to give the speeches, but they didn't know this until they exited the scanner.

While they were lying in the fMRI machine preparing their speeches, students' heart rates were continuously monitored and they were asked to report, about every twenty seconds, how much anxiety they were feeling at that time. Not surprisingly, Wager and his research team found that the anticipation of giving a speech changed people's heart rates and reported anxiety levels and that activation in areas of the prefrontal cortex explained the link between speech anticipation and anxiety (especially so for those who viewed the speech preparation task as most anxiety-provoking in the first place). When getting ready to give a speech, the more activity in these prefrontal regions, the more anxious these people were.

One interpretation of Wager's findings is that the more people nervous about speech giving dwelled on what others would think—the more they anticipated the panel of experts' reactions—the more anxious they became. Keep in mind that all of these brain changes occurred before students had done anything. This suggests that the anticipation of an event, and specifically anticipation of others judging you, is enough to put pressure on you before you have even arrived at the performance stage. If the end result is choking under pressure, then we have somewhat of a recursive cycle on our hands. You worry about how others will judge you, which may lead to poor performance, which leads to more worry the next time you are in a public speaking situation, and so on.

Performance anxieties that stem from how others may judge you are of course not limited to public speaking. Case in point is a famous dialogue aired in a third-season *Seinfeld* episode:

> George: I don't like when a woman says, "Make love to me." It's intimidating. The last time a woman said that to me, I wound up apologizing to her.

Jerry: Really?

George: That's a lot of pressure. "Make love to me." What am I, in the circus? What if I can't deliver?

Jerry: Oh, come on.

George: I can't perform under pressure. That's why I never play anything for money. I choke. I could choke tonight. And she works in my office, can you imagine? She goes around telling everyone what happened? Maybe I should cancel. I have a very bad feeling about this.

Jerry: George, you're thinking too much.

George: I know, I know, I can't stop it!

High expectations for success and the possibility that you will be evaluated poorly can lead to disastrous consequences in the boardroom and in the bedroom. As we have seen from Tor Wager's work, even when Columbia University students are merely preparing to give a speech, a variety of brain and body reactions occur that can send people down a path to failure. These sorts of anticipation effects likely happen in the ultimate performance situation, sex, as well.

A friend of mine told me about a long-distance relationship he once had with a woman in college. She was his first true love, but unfortunately they lived quite a distance apart from each other and were only able to spend one weekend a month together. The anticipation of their encounters, he told me, really backfired. As soon as they were under the covers together, he couldn't perform. The fact that they only had one shot a month got him thinking about it ahead of time and worrying about the outcome, and so on.

Dr. Robert J. Filewich, executive director for the Center for Behavior Therapy in White Plains, New York, and a clinical psychologist specializing in anxiety disorders, says that "performance anxiety, in sexual terms and with sexual problems, is where a person has an anticipation of some sort of problem occurring in the sexual act. As a consequence of that, they develop a sense of anxiety which translates into an inability to become erect or an inability to go ahead and have sex for a certain duration before they actually achieve orgasm or premature ejaculation."[17] Anticipation can be a problem.

What do you do about this type of performance flub? There are several pieces of advice that researchers and physicians have offered.[18] First, it is important to remember that some things are a reflex response and not necessarily under your conscious control. Thus, just as thinking step-by-step about the components of a golf swing can cause a golfer to miss a simple putt, concentrating and focusing on your performance in the sack can be detrimental. As Jerry Seinfeld said, "George, you're thinking too much!" Second, focusing on the negative can lead to feeling like you have no control, similar to what we saw in chapter 8 with the Canadian Olympic swimmers. After watching a failed race, these swimmers showed signs of learned helplessness—a loss of a sense of control and a tendency to give up because of it. The same thing can happen in the bedroom after a failed engagement. Focusing on the positive will help ensure that a reactive cycle does not occur.

Many of the same factors that lead to poor performances in public speaking can also be at work in the bedroom. Relationship factors, including problems in communication, conflict, and quality, can detract from enjoyment and performance for both men and women. Stress from other aspects of life can seep in and distract folks from the task at hand. In addition, spouses can be supportive in the sack and increase the probability of success, or be unsupportive and, just like having them around before a big speech, this lack of support can backfire and actually cause more problems than if you were alone.

LOSING CONTROL OF YOUR PREFRONTAL CORTEX

"In the heat of battle I let my passion and emotion get the better of me and as a result handled the situation poorly." That statement was released by tennis player Serena Williams after an outburst at a line judge cost her the match and an early exit from the 2009 U.S. Open.

Serena Williams was playing Kim Clijsters in a much-anticipated semifinal match. The first set had gone to Clijsters, 6-4. It was a heated opening set and Williams had already been issued a warning and fined five hundred dollars for smashing her racquet in frustration. When she was one point shy of match point in the second set, the line judge called a foot fault on Serena. Foot faults are rarely called in this situation and it no doubt surprised Serena to lose a point when there were possibly only two remaining. Instead of taking her next serve, Serena Williams approached the official who had made the call; she was waving the ball and screaming. This outburst cost her ten thousand dollars, the final point, and the match.

Of course, if you believe that all publicity is good publicity, then the bright side of the entire situation is that the screaming scene made the "most viewed" page of YouTube that weekend. It's not clear exactly what words were uttered, but on my watch of the video, I did catch that Serena threatened to shove the ball down the judge's throat.

When people are in stressful, high-stakes situations their ability to inhibit their emotions and unwanted behaviors is impaired. A major component of working-memory is inhibition, which helps us keep what we want in mind and what we don't want out. It also helps us control our thoughts and behavior. When the stress is on, working-memory and the prefrontal cortex can be compromised and our inhibition is one of the first things to go. When you lose your temper or say things you shouldn't, it's often a sign that your prefrontal cortex isn't able to keep the emotional centers of your brain under wraps.

The prefrontal cortex is also the seat of our ability to reappraise a situation or event. Reinterpreting a coworker's angry behavior so that you can better understand the cause of the behavior, reappraisal is one of the main cognitive tools we use to reflect on what others do and change our own emotional responses accordingly.[19] Usually working-memory drives this reappraisal process and helps us choose how to act rather than just impulsively react. When our prefrontal cortex is not functioning on all cylinders, however, a heated negotiation or even a talk with a coworker or employee can go awry. This is certainly true for mothers dealing with the challenging behavior of an angry or nonresponsive child. Researchers have shown that when parents' working-memory is at its lowest, their tendency to react negatively to their oppositional children is highest. Rather than reason with a disobedient and challenging child in a nonemotional way, parents with low working-memory tend to meet anger with anger, which doesn't usually lead to a positive outcome for either parent or child.[20]

Teenagers have the propensity to bring emotions to the table when they are best left off. Because the prefrontal cortex is still developing in adolescents, they often have a hard time keeping the emotional areas of the brain in check. When young teens and adults in their mid-twenties to mid-thirties were asked to look at pictures of emotional faces presented on a computer screen, teens tended to exhibit greater activation in the amygdala than adults did along with orbitofrontal cortex and anterior cingulate cortex, part of the brain systems involved in fear and appraisal of emotional situations.[21] When people were asked to switch their attention between an emotional component of the face (such as thinking about how the face makes them feel afraid) and a nonemotional feature (how far apart the eyes are spaced), the adults' brains were much better at doing this. When needed, adults seem to be better able turn off their emotional brain areas or at least keep it in check in a way the teens are not able to do.

When adults are under pressure, however, everything changes. The prefrontal cortex stops working the way it should, which can result in overattention to performance, a lack of cognitive horsepower devoted to the task at hand, or an emotional outburst that seems more typical of a teenager. As teens grow older, the prefrontal cortex develops and people are better able to modulate their reactions. But, under stress, this control can go out the window. Just think about French soccer player Zinedine Zidane and his infamous head butt in the 2006 World Cup finals. After Italian defender Marco Materazzi and he exchanged heated words, Zidane's prefrontal cortex was likely working hard to restrain an emotional outburst. But with the stress of the world championship on the line, this inhibition did not materialize and instead of walking away from Materazzi, Zidane rammed his head into Materazzi's chest, sending the Italian to the ground and Zidane from the game. Handling yourself appropriately under pressure involves recognizing when your prefrontal cortex is most likely to resemble that of a teenage brain and applying effective techniques to deal with the regression.

After the foot fault call at the 2009 open, Serena Williams actually began to prepare for her next serve before she stopped and walked up to the line judge. There is a good chance Williams would have been better off had she walked away for a moment to begin with. As we talked about earlier in this book, taking a step back can help people see a problem from a new perspective and also prevent emotions from taking over. Moreover, merely educating people that their prefrontal cortex will be compromised under stress and they may be more reactive in nature can actually make them less likely to react poorly.

A few weeks after getting back from my talk with the VPs at Sundance, I received an e-mail from John, an executive who had been in the audience during my presentation who had told a story about his daughter's math experience after I had mentioned that it was those people highest in cognitive horsepower who were most likely to choke under pressure. John's daughter seemed to fit my choking description to a T: she consistently got the best marks in her class on homework assignments and short quizzes, but panicked on important tests and often bombed them because of it.

The reason for the e-mail, John explained, was that he wanted to tell me of an experience he had recently had on the golf course. While playing eighteen holes, in front of the watchful gazes of his foursome, he had lost his cool and shanked several important shots that cost him the match. The entire situation seemed quite similar to what his daughter experienced on big tests. John wanted to know whether his flubbed performance on the links was one and the same with what happened to his daughter in the classroom.

All high-pressure situations can induce similar brain and body reactions. Our heart rate goes up, our adrenaline kicks in, and our minds start to race—often with worries. When the worries start, if we are doing something that demands a heavy dose of cognitive horsepower our performance will suffer. But worries alone are not usually the problem when we're faced with a simple putt or action that doesn't require a lot of thought. Highly practiced skills don't draw heavily on the brainpower that worries involve. But when the stakes are high, we try to control what we are doing, which can backfire. Bringing highly practiced routines back into conscious control disrupts them—we suffer paralysis by analysis.

As an example, think back to Chief Justice John Roberts fumbling Obama's presidential oath in front of the entire free world. There is no doubt that Roberts had practiced his duties many times, but when the pressure was on, he fumbled the well-learned set of lines, likely because Roberts was devoting too much working-memory to monitoring the words he knew by heart.

So, yes, the pressures we face in the boardroom, classroom, and playing field exert common effects on the mind and body. Under pressure, worries flood the brain and, as a result, we try to manage what we are doing. If we are engaged in an activity that demands a lot of prefrontal brain power, such as taking a difficult test or making demanding decisions, worries alone can send performance awry. Worries can also prevent us from being able to use our working-memory to inhibit unwanted thoughts or behavior. Yet if instead we are performing a skill that runs outside conscious awareness, like John's highly practiced putt, we may screw up because our cognitive horsepower (or what horsepower that remains after worries have done their damage) is leveraged to control performance in a counterproductive way.

High-stakes situations can affect performance whether you have a pencil or golf club in your hand, but the reason you choke depends on the characteristics of the task and on you yourself. That's why different solutions work for the different choking that occurs in the classroom, boardroom, sports field, and orchestra pit.

For instance, singing a song to create a mild distraction might increase your likelihood of nailing an easy three-foot putt to win the tournament, but this same distraction could be devastating for a student combating pressure during the SAT. When you are trying to answer a difficult question where logic and reasoning are a must—in a test or in a business meeting—attention to performance is a good thing. Getting students to talk aloud while solving math problems, for example, gets them focused on the task at hand and limits the possibility of worries and other distractions creeping in.[22]

Likewise, pushing a field goal kicker to speed up his pre-shot routine might prevent him from overthinking his game-winning kick, but speeding up isn't a good idea for a student trying to solve a difficult physics problem, which might make him or her overlook critical equation details.

Sometimes you may need several different pressure-fighting strategies at once—as when you find yourself delivering an important presentation that you've practiced to perfection while at the same time you have to field difficult questions on the fly. To succeed in this pressure-filled situation, you will not only have to combat worries, you will also have to make sure you don't exert too much control over your well-practiced speech routine. Understanding why different high-pressure situations can derail performance allows you to pick the right strategy to prevent choking.

Here are some tips that you can use to prevent choking in the business world, whether you are trying to avoid thinking too much and derailing your practice or need all the cognitive horsepower you can muster to think effectively on the fly.

Prevent the Choke

Be a memory guide. In the interview, give a schema at the outset that will help the interviewer encode your positive attributes. Start off by telling your interviewer why you are the best person for the job.

Subtle mimicry can help create positive affect and liking in interview situations by putting your and your interviewers' brain on the same page.

Think about what you want to say, not what you don't want to say, because when you try not to think or do something, it may be more likely to occur.

Practice making a fool out of yourself in a comedy, acting, or improvisation class. That way, when you are faced with making a speech you will be less likely to worry about what might happen if you stumble, because you will have already experienced it.

Know what you know. If you have memorized the introduction to your speech or what you are going to say in its entirety, just go with it and try not to think too much about every word. If not, pause before key transitions to allow yourself time to regroup.

Write it out. Research shows that writing about worries and stressful events in your life can help increase working-memory and may prevent other parts of your life (spouse, kids, house) from creeping in and distracting you under stress.

Think about the journey, not the outcome. Being so focused on failing or the monumental goals you are trying to achieve may prevent you from making the small steps forward needed to succeed.

Remind yourself that you have the background to succeed and that you are in control of the situation. This can be the confidence boost you need to ace your pitch.

Prepare well, but don't anticipate too much. Often it is the stress you give yourself worrying about the "what-ifs" that leads to failure when the pressure is on—a self-fulfilling prophecy.

EPILOGUE

ROMA NEVER FORGETS

1984 UEFA CHAMPIONS LEAGUE FINALS: ROMA 1, LIVERPOOL 1: LIVERPOOL WINS 4-2 IN PENALTIES.

On May 30, 1984, two of the world's top soccer clubs—Roma and Liverpool—met at Stadio Olimpico in Rome, Italy, to play in the finals of the Champions League. Bringing together the top clubs in Europe, the Champions League represents the best of the best in soccer.

There were 69,693 fans on hand that day to watch the finals. But this number is only a small drop in the bucket of the millions of people around the world glued to television and radio sets, who had turned in for the big game. Liverpool had already won the Champions League three times previously—in both 1977 and 1978 and again in 1981—so they were somewhat used to the pressure of this important game. Roma, on the other hand, had never taken home what is arguably the most prestigious trophy in club soccer. In fact, Roma had never competed in the Champions League finals before and hasn't done so since.

It was an exciting game and Liverpool scored early on. However, Roma answered back when Roberto Pruzzo connected with a beautiful cross from Bruno Conti, heading the ball into goal. Regulation play ended in a 1-1 tie and extra time didn't change this. Penalty shots would decide the game.

Liverpool took the first kick and actually missed, while Roma converted their penalty into a goal. But Roma's lead would stop there. In the second round, Liverpool scored and Conti, who had set up Roma's only goal during regulation play, missed his shot. The score was tied 1-1. Both teams made their third-round goals and, in round four Liverpool's Ian Rush made his shot as well—advantage Liverpool. Francesco Graziani stepped up to the ball next. Graziani was a leading scorer for Roma and by all accounts should have been able to score effortlessly. But Graziani's kick hit the crossbar and sailed high over the goal instead, landing in the grandstands. Liverpool capitalized on their last attempt, won the penalty kick lottery, and the European Cup for the fourth time.

At first glance you might have expected that everything was in place for Roma to take home the Champions League trophy. After all, Roma was playing at home in front of supporters and fans and both Conti and Graziani were superstar players, chosen to take penalty kicks because of their consistent ability to hit the back of the net. Yet once you know the specific factors that can dial up the stress, it becomes clear that this penalty round wasn't conducive to Roma's success.

Ironically, performing in front of supportive audiences in a decisive game—in front of a home-field crowd—can lead to disaster. Players become self-conscious about themselves and their movements. When a professional footballer takes a kick he has practiced hundreds of thousands of times in the past, this type of attention to detail can derail his performance.[1] This home-field disadvantage has been documented in critical games and tournaments in professional baseball, basketball, ice hockey, and golf.[2] There is no doubt that Roma experienced the pressures of being at home that day in the Champions League finals.

Soccer players who are publicly esteemed superstars, such as Bruno Conti and Francesco Graziani, are also more prone to miss penalty shots than those players who are not stars, research has shown.[3] When the weight of the expectations of fans, supports, sponsors, teammates, and coaches is on your shoulders, performing at your best becomes harder than you might think.

Finally, being faced with the pressure of scoring when your team trails by only one point often leads to suboptimal performance. In the NBA, for instance, researchers have found that the probability of a player converting a free throw in the final minutes of a game when his team is trailing by only one point is about 7 percent lower than the player's season-average free-throw performance.[4] When Graziani stepped up to the ball with Roma trailing 3-2, the deck was stacked against him.

Coaches, team leaders, and managers like to say that "pressure makes diamonds." There is no denying that a spectacular performance in a high-stakes situation can take a once-unknown individual and make him or her a household name. But folks rarely consider that these factors can also lead to suboptimal performance. When you are on the home field, carrying the weight of superstar expectations, and playing for a trailing team, even if you are the world's best you can choke under the pressure.

In this book we have looked to research that documents the situations under which choking will occur; knowledge that can help performers prepare for what lies ahead before they even step onto the field or stage, or into the boardroom or classroom. We have seen that although all high-stress situations have similarities, there are differences in how pressure affects the brain and body to induce choking. The tricky issue is that there are certain activities in which less working-memory and input from your prefrontal cortex are beneficial to performance and certain activities in which you need all the cognitive horsepower you can muster to succeed. Because being in a pressure-filled situation can push people to think about what they are doing in precisely the ways they should avoid, deciding how to handle stress involves recognizing what happens when you fail in your own performance endeavors and knowing what appropriate pressure-management techniques to apply.

Of course, the science of human performance has yet to get a full handle on choking. With the continuing advancement of brain-imaging technology, however, I have no doubt that we will unlock all the keys to success and failure in the years to come. At some point we may be able to strap a small fMRI onto a golfer as he or she actually plays a round of golf so that we can monitor exactly what the brain is doing during triumph and defeat. Indeed, we may even be able to scan an entire team simultaneously so that we can not only look at how an individual's brain communicates with itself to bring about success but also how coherence among a network of different brains leads to championship performance. Our workforce, the military, academia, and sports teams are dominated by people working as units to succeed. Understanding how groups thrive and dive under pressure will help to improve everyone's performance.

In reality, broken world records don't count unless someone is watching, a perfect score on an SAT means nothing if it is a practice test, and a spectacular solo is lost in an empty auditorium. Yet when choking under pressure happens, people don't forget. Yes, Conti and Graziani are known as great soccer players, but if you ask any Roman in his mid-thirties on up to talk about that 1984 Champions League final, you can almost see the tears well up in his eyes.

Being good at what you do necessitates being able to perform well when it counts the most. Thus knowing what factors produce the highest levels of pressure, how to practice your skill so that you are ready for whatever comes your way, and understanding how to handle the pressure when it inevitably does arrive, can make all the difference between moving up and moving on. I hope you can use the science behind choking to combat pressure's negative effects.

Of course, even when you find yourself armed with information about how to ensure success under stress, it's important to remember that—sometimes—it's better not to step into competition in the first place. In 1974, Canadian Progressive Conservative leader Robert Stanfield played a game of catch in front of several press photographers during a stop on his campaign trail. The next day, a picture of a bewildered-looking Stanfield dropping the football made the front page of the *Globe & Mail.* That fumble is rumored to have cost Stanfield the election. When all eyes are on you, you can miss even simple lobs. Politicians usually spend hours and hours perfecting their speeches and even practice giving them in front of others, but this practice doesn't usually extend to their football-receiving skills. If you are not in a position to practice making the catch when the pressure is on, it may be better to avoid the game to begin with. Good luck!

ACKNOWLEDGMENTS

This book would not have been possible without the support and guidance of so many people. First and foremost, I would like to thank my family—my mom, Ellen, my late father, Steve, and my brother, Mark—who have always encouraged me to be the best I can be. This also includes my extended parents, Judy and Dave Stein. I couldn't imagine a better set of cheerleaders. I would also like to thank my Ph.D. advisers, Deb Feltz and Tom Carr. Without your help and direction, I wouldn't be where I am today. And, thanks to my students, postdocs, and colleagues who were most gracious in letting me bounce ideas off them during the writing process.

Thanks to all those people who read and gave comments on earlier versions of this book. Your input was most valuable. Thanks also to Dario for your unwavering confidence in this book and your willingness to drop whatever you were doing to provide me with instantaneous feedback. Finally, thanks to my editor, Leslie Meredith, and my agents, Dan O'Connell and Wendy Strothman, for their patience, support, and guidance.

NOTES

1. THE CURSE OF EXPERTISE

1. P. Hinds, "The curse of expertise: The effects of expertise and debiasing methods on predictions of novice performance," *Journal of Experimental Psychology: Applied, 5* (1999), 205–21.
2. I am using the term *explicit memory* here to refer to both explicit memory that governs our ability to remember and work with information in the short term (what I refer to as working memory) and explicit long-term memory that is often divided into semantic memory (our memory for particular facts, such as "dogs often bark") and episodic memory (our memory for autobiographical experiences, such as "the first time you met your spouse"). For a memory review, see G. Radvansky, *Human Memory* (New York: Pearson, 2006).
3. S. Corkin, D. G. Amaral, R. G. Gonzalez, K. A. Johnson, and B. T. Hyman, "H. M.'s Medial Temporal Lobe Lesion: Findings from Magnetic Resonance Imaging," *Journal of Neuroscience, 17* 1997, 3964–79. For a more general overview of H. M., see E. E. Smith and S. M. Kosslyn, *Cognitive Psychology: Mind and Brain* (Upper Saddle River, N.J.: Prentice Hall, 2007).
4. M. K. Smith, W. B. Wood, et al., "Why Peer Discussion Improves Student Performance on In-Class Concept Questions," *Science, 323* (2009), 122–24.
5. S. L. Beilock and M. S. DeCaro, "From poor performance to success under stress: Working memory, strategy selection, and mathematical problem solving under pressure," *Journal of Experimental Psychology: Learning, Memory, & Cognition, 33* (2007), 983–98.
6. For a brief general overview of working memory, see R. W. Engle, "Working memory capacity as executive attention," *Current Directions in Psychological Science, 11* (2002), 19–23. For links between working memory and fluid intelligence, see M. J. Kane, D. Z. Hambrick, and A. R. A. Conway, "Working memory capacity and fluid intelligence are strongly related constructs: Comment on Ackerman, Beier, and Boyle," *Psychological Bulletin, 131* (2005), 66–71, and K. Oberauer, R. Schulze, O. Wilhelm, and H.-M. Süss, "Working memory and intelligence—Their correlation and their relation: Comment on Ackerman, Beier, and Boyle," *Psychological Bulletin, 131* (2005), 61–65, and, H.-M. Süss, K. Oberauer, W.W. Wittmann, O. Wilhelm, and R. Schulze, "Working-memory capacity explains reasoning ability—and a little bit more," *Intelligence 30* (2002), 261–88.
7. Redrawn with permission from BrainVoyger.
8. For a review of complex working-memory span tasks, see A. R. A. Conway, et al., "Working memory span tasks: A methodological review and user's guide," *Psychonomic Bulletin & Review, 12* (2005), 769–86.
9. M. T. Chi, P. J. Feltovitch, and R. Glaser, "Categorization and representation of phys-

ics problems by experts and novices," *Cognitive Science, 5* (1981), 121–52. Note that the studies conducted by Chi et al. involved not only actual physics problem solving but an examination of how novice and experts sorted physics problems into distinct categories (e.g., by underlying physics principle versus surface feature) as a way to assess problem-solving differences across people with more or less physics experience.

10. R. R. D. Oudejans, "Reality based practice under pressure improves handgun shooting performance of police officers," *Ergonomics, 51* (2008), 261–73.
11. J. Milton, A. Solodkin, P. Hlustik, and S. L. Small, "The mind of expert motor performance is cool and focused," *Neuroimage, 35* (2007), 804–13.

2. TRAINING SUCCESS

1. To learn more about MRI and fMRI, see S. A. Huettel, A. W. Song, and G. McCarthy, *Functional Magnetic Resonance Imaging* (Sunderland, MA: Sinauer Associates, 2004), and M. S. Gazzaniga, R. B. Ivry, and G. R. Mangun, *Cognitive Neuroscience: The Biology of the Mind* (New York: Norton, 2002).
2. *Saturday Matters with Sue Lawley,* BBC Television, October 1980, as quoted in R. Masters and J. Maxwell, "The theory of reinvestment," *International Review of Sport and Exercise Psychology, 1* (2009), 160–83.
3. D. P. McCabe and A. D. Castel, "Seeing is believing: The effect of brain images on judgments of scientific reasoning," *Cognition, 107* (2008), 343–52.
4. W. F. Helsen and J. L. Starkes, "A multidimensional approach to skilled perception and performance in sport," *Applied Cognitive Psychology, 13* (1999), 1–27.
5. W. F. Helsen, J. Van Winckel, and A. M. Williams, "The relative age effect in youth soccer across Europe," *Journal of Sports Sciences, 23* (2005), 629–36.
6. J. Bisanz, F. Morrison, and M. Dunn, "Effects of age and schooling on the acquisition of elementary quantitative skills," *Developmental Psychology, 31* (1995), 221–36.
7. J. Cote, D. J. MacDonald, J. Baker, and B. Abernethy "When 'where' is more important than 'when': Birthplace and birthdate effects on the achievement of sporting expertise," *Journal of Sport Sciences, 24* (2006), 1065–73. Note that in Canada, the smallest rural areas of less than one thousand people were excluded from the analyses.
8. A. D. De Groot, *Thought and Choice in Chess* (The Hague: Mouton, 1965).
9. W. G. Chase and H. A. Simon, "Perception in chess," *Cognitive Psychology, 4* (1973), 55–81.
10. K. A. Ericsson and P. G. Polson, "An experimental analysis of the mechanisms of a memory skill," *Journal of Experimental Psychology: Learning, Memory, & Cognition, 14* (1988), 305–16.
11. Maguire et al., "Navigation-related structural change in the hippocampi of taxi drivers," *Proceedings of the National Academy of Sciences, 97* (2000), 4398–4403.
12. B. Draganski, "Changes in grey matter induced by training: Newly honed juggling skills show up as a transient feature on a brain-imaging scan," *Nature, 427* (2004), 311–12.
13. For a review, see T. F. Münte, E. Altenmüller, and L. Jäncke, "The musician's brain as a model of neuroplasticity," *Nature Reviews Neuroscience, 3* (2002), 473–78.
14. For a review, see A. E. Hernandez and P. Li, "Age of acquisition: Its neural and computational mechanisms," *Psychological Bulletin, 133* (2007), 638–50.

15. This is the Movement Specific Reinvestment Scale. Reprinted with permission. See R. S. W. Masters, F. F. Eves, and J. Maxwell, "Development of a movement specific reinvestment scale," in T. Morris et al., eds., *Proceedings of the ISSP 11th World Congress of Sport Psychology,* Sydney, Australia (2005). For the original Reinvestment Scale used with the squash and tennis players mentioned below, see chapter 7 and R. S. W. Masters, R. C. J. Polman, and N. V. Hammond, " 'Reinvestment': A dimension of personality implicated in skill breakdown under pressure," *Journal of Personality and Individual Differences, 14:5* (1993), 655–66.
16. For a general overview, see R. Masters and J. Maxwell, "The theory of reinvestment," *International Review of Sport and Exercise Psychology, 1* (2008), 160–83.
17. See K. Yarrow, P. Brown, and J. W. Krakauer, "Inside the brain of an elite athlete: The neural processes that support high achievement in sports," *Nature Reviews Neuroscience, 10* (2009), 585–96.

3. LESS CAN BE MORE

1. J. Wiley, "Expertise as mental set: The effects of domain knowledge in creativity," *Memory & Cognition, 26* (1998), 716–730. See also, T. Ricks, K. J. Turley-Ames, and J. Wiley, "Effects of working memory capacity on mental set due to domain knowledge," *Memory and Cognition 35* (2007), 1456–62.
2. Copyright © (2009) by the American Psychological Association. Adapted with permission, W. W. Maddux, and A. D. Galinsky, "Cultural borders and mental barriers: The relationship between living abroad and creativity," *Journal of Personality and Social Psychology, 96* (2009), 1047–61. The use of APA information does not imply endorsement by APA.
3. For a review, see S. L. Thompson-Schill, M. Ramscar, and E. G. Chrysikou, "Cognition without control: When a little frontal lobe goes a long way," *Current Directions in Psychological Science, 18* (2009), 259–263. See also, T.P. German and M.A. Defeyter, "Immunity to functional fixedness in young children," *Psychonomic Bulletin & Review, 7* (2000), 707–12.
4. S. L. Beilock and M. S. DeCaro, "From poor performance to success under stress: Working memory, strategy selection, and mathematical problem solving under pressure, *Journal of Experimental Psychology: Learning, Memory, & Cognition, 33* (2007), 983–98.
5. A. R. A. Conway, N. Cowan, and M. F. Bunting, "The cocktail party phenomenon revisited: The importance of working memory capacity," *Psychonomic Bulletin and Review, 8* (2001), 331–35.
6. A. W. Kersten and J. L. Earles, "Less really is more for adults learning a miniature artificial language," *Journal of Memory and Language, 44* (2001), 250–273. For a general review, see also, E. L. Newport, "Maturational constraints on language learning," *Cognitive Science, 14* (1990), 11–28.
7. C. Reverberi, A. Toraldo, S. D'Agostini, and M. Skrap, "Better without (lateral) frontal cortex? Insight problems solved by frontal patients," *Brain, 128* (2005), 2882–2890.
8. B. P. Cochran, J. L. McDonald, and S. J. Parault, "Too smart for their own good: The disadvantage of a superior processing capacity for adult language learners," *Journal of Memory and Language, 41* (1999), 30–58.
9. S. L. Beilock, T. H. Carr, C. MacMahon, and J. L. Starkes, "When paying attention becomes counterproductive: Impact of divided versus skill-focused attention on nov-

ice and experienced performance of sensorimotor skills. *Journal of Experimental Psychology: Applied, 8* (2002), 6–16.

10. J. M. Ellenbogen, P. T. Hu, J. D. Payne, D. Titone, and M. P. Walker, "Human relational memory requires time and sleep," *Proceedings of the National Academy of Sciences, USA, 104* (2007), 7723–28.
11. N. P. Friedman, et al., "Individual differences in executive functions are almost entirely genetic in origin," *Journal of Experimental Psychology: General, 137* (2008), 201–25.
12. J. Ward, *The Student's Guide to Cognitive Neuroscience*, 2nd ed. (London: Psychology Press, 2010).
13. T. Klingberg, H. Forssberg, and H. Westerberg, "Training of working memory in children with ADHD," *Journal of Clinical and Experimental Neuropsychology, 24* (2002), 781–791. See also, T. Klingberg, et al., "Computerized training of working memory in children with ADHD—a randomized, controlled trial," *Journal of the American Academy of Child Adolescent Psychiatry, 44* (2005), 177–186. For a general review, see T. Klingberg, *The Overflowing Brain: Information Overload and the Limits of Working Memory* (New York: Oxford University Press, 2009).
14. M. I. Posner, and M. K. Rothbart, "Influencing brain networks: implications for education," *Trends in Cognitive Sciences, 9* (2005), 99–103. For a general review, see, M. K. Rothbart and M. I. Posner, *Educating the Human brain* (Washington, D.C.: American Psychological Association, 2006).
15. P. J. Olesen, H. Westerberg, and T. Klingberg, "Increased prefrontal and parietal activity after training of working memory," *Nature Neuroscience, 7* (2003), *75–79*.
16. C. S. Green, and D. Bavelier, "Action video games modifies visual selective attention," *Nature, 423* (2003), 534–37.
17. D. Gopher, M. Weil, and T. Bareket, "Transfer of skill from a computer game trainer to flight," *Human Factors, 36* (1994), 1–19. See also, "Sharp Brains" interview with Daniel Gopher, November 2, 2006: http://www.sharpbrains.com/blog/2006/11/02/cognitive-simulations-for-basketball-game-intelligence-interview-with-prof-daniel-gopher/.
18. H. A. White, and P. Shah, "Uninhibited imaginations: Creativity in adults with Attention-Deficit/Hyperactivity Disorder." *Personality and Individual Differences, 40* (2006), 1121–31.

4. BRAIN DIFFERENCES BETWEEN THE SEXES

1. L. H. Summers, Remarks at NBER Conference on Diversifying the Science & Engineering Workforce, Cambridge, Mass., January, 14, 2005.
2. C. P. Benbow and J. C. Stanley, "Sex differences in mathematical reasoning ability: More facts," *Science, 222* (1983), 1029–30.
3. P. L. Ackerman, "Cognitive sex differences and mathematics and science achievement," *American Psychologist, 61* (2006), 722–23.
4. For a review, see S. A. Shields, Functionalism, Darwinism, and the psychology of women: A study in social myth," *American Psychologist* (1975), 739–54.
5. "Summers' remarks on women draw fire," *Boston Globe,* January 17, 2005.
6. R. Herrnstain and C. Murray, *The Bell Curve: Intelligence and Class Structure in American Life* (New York: Free Press, 1994).

7. C. P. Benbow, and J. C. Stanley, "Sex differences in mathematical reasoning ability: More facts," *Science, 222* (1983), 1029–30. See also C. P. Benbow and J. C. Stanley, "Sex differences in mathematical ability: Fact or artifact?" *Science, 210* (1980), 1262–64.
8. L. Brody and C. Mills, "Talent search research: What have we learned?" *High Ability Studies, 16,* 97–111. See also R. Monastersky, "Studies show biological differences in how boys and girls learn about math, but social factors play a big role too," *Chronicle of Higher Education, 51* (March 4, 2005).
9. J. S. Hyde and J. E. Mertz, "Gender, culture, and mathematics performance, *PNAS, 106* (2009), 8801–7. See also Title IX, Education Amendments of 1972, United States Department of Labor, http://www.dol.gov/oasam/regs/statutes/titleIX.htm.
10. G. Ellison and A. Swanson, "The gender gap in secondary school mathematics at high achievement levels: Evidence from the American Mathematics Competitions, NBER Working Paper No. 15238, August 2009.
11. S. M. Rivera, A. L. Reiss, M. A. Eckert, and V. Menon, "Developmental changes in mental arithmetic: Evidence for increased specialization in the left inferior parietal cortex," *Cerebral Cortex, 15* (2005), 1779–90.
12. For a review, see D. F. Halpern et al., "The science of sex differences in science and mathematics. *Psychological Science in the Public Interest, 8* (2007), 1–51.
13. S. M. Resnick, S. A. Berenbaum, I. I. Gottesman, and T. J. Bouchard, "Early hormonal influences on cognitive functioning in congenital adrenal hyperplasia," *Developmental Psychology, 22* (1986), 191–98. See also M. A. Maloufa et al., "Cognitive outcome in adult women affected by congenital adrenal hyperplasia due to 21-hydroxylase deficiency," *Hormone Research in Pediatrics, 65* (2006), 142–50.
14. R. Monastersky, "Studies show biological differences in how boys and girls learn about math, but social factors play a big role too," *Chronicle of Higher Education, 51,* March 4, 2005.
15. For a review, see E. S. Spelke, "Sex differences in intrinsic aptitude for mathematics and science? A critical review," *American Psychologist, 60* (2005), 950–58.
16. A. Gallagher and R. DeLisi, "Gender differences in Scholastic Aptitude Test–Mathematics problem solving among high-ability students," *Journal of Educational Psychology, 86* (1994), 204–11.
17. E. Fennema, T. Carpenter, V. Jacobs, M. Franke, and L. Levi, "A longitudinal study of gender differences in young children's mathematical thinking," *Educational Researcher, 27* (1996), 33–43.
18. L. Butler, (1999) "Gender differences in children's arithmetical problem solving procedures," unpublished M.A. thesis, University of California at Los Angeles, as cited in Association for Women in Mathematics president Cathy Kessel's talk at the MER-AWM Session at the 2005 Joint Mathematics Meetings.
19. Richard C. Atkinson, "Let's step back from the SAT I," *San Jose Mercury News,* February 23, 2001, http://www.ucop.edu/pres/comments/satmerc.html.
20. S. J. Spencer, C. M. Steele, and D. M. Quinn, "Stereotype threat and women's math performance," *Journal of Experimental Social Psychology, 35* (1999), 4–28.
21. For a review, see C. M. Steele, "A threat in the air: How stereotypes shape intellectual identity and performance," *American Psychologist, 52* (1997), 613–29, and T. Schmader, M. Johns, and C. Forbes, "An integrated process model of stereotype threat effects on performance," *Psychological Review, 115* (2008), 336–56.
22. A. C. Krendl, J. A. Richeson, W. M. Kelley, and T. F. Heatherton, "The negative consequences of threat: A functional magnetic resonance imaging investigation of the neu-

ral mechanisms underlying women's underperformance in math," *Psychological Science, 19* (2008), 168–75. See also S. L. Beilock, "Math performance in stressful situations," *Current Directions in Psychological Science, 17* (2008), 339–43.

23. S. C. Levine et al., "Socioeconomic status modifies the sex difference in spatial skill," *Psychological Science, 16* (2005), 841–45.
24. T. G. Thurstone, *PMA readiness level* (Chicago: Science Research Associates, 1974). Redrawn with permission.
25. In 2001, 90 percent of all Legos sold in the United States were intended for boys as reported in the *Wall Street Journal* on June 6, 2002, "Mattel sees untapped market for blocks: Little girls." Moreover, as reported in the *Wall Street Journal,* December 24, 2009, Legos has still failed to crack the girls market: http://online.wsj.com/article/SB10001424052748704254604574613791179449708.html.
26. J. M. Hassett, E. R. Siebert, and K. Wallen, "Sex differences in rhesus monkey toy preferences parallel those of children," *Hormones & Behavior, 54* (2008), 359–64.
27. K. Wallen, "Hormonal influences on sexually differentiated behavior in nonhuman primates," *Frontiers in Neuroendocrinology, 26* (2005), 7–26.
28. J. Connellan et al., "Sex differences in human neonatal social perception," *Infant Behavior & Development, 23* (2000), 113–18.
29. For a review, see E. S. Spelke, "Sex Differences in intrinsic aptitude for mathematics and science? A critical review," *American Psychologist, 60* (2005), 950–58.
30. L. Guiso, F. Monte, P. Sapienza, and L. Zingales, "Culture, gender, and math," *Science, 320* (2008), 1164–65.
31. "Mattel says it erred; Teen Talk Barbie turns silent on math, *New York Times*, October 21, 1992, http://www.nytimes.com/1992/10/21/business/company-news-mattel-says-it-erred-teen-talk-barbie-turns-silent-on-math.html.
32. D. N. Figlio, "Why Barbie says 'Math is Hard,'" Working Paper, University of Florida, December 2005.
33. To read more about gender differences and names, see S. Pinker, *The Stuff of Thought: Language as a Window into Human Nature* (New York: Viking, 2007).
34. M. C. Murphy, C. M. Steele, and J. J. Gross, "Signaling threat: how situational cues affect women in math, science, and engineering settings," *Psychological Science, 18* (2007), 879–85.
35. C. M. Steele, "A threat in the air: How stereotypes shape intellectual identity and performance," *American Psychologist, 6* (1997), 613–29.
36. P. Tyre, *The Trouble with Boys: A Surprising Report Card on Our Sons, Their Problems at School, and What Parents and Educators Must Do* (New York: Three Rivers Press, 2008).
37. College Board, *Summary Reports: 2007: National Report,* http://www.collegeboard.com/student/testing/ap/exgrd_sum/2007.html.

5. BOMBING THE TEST

1. R. Hembree, "The nature, effects, and relief of mathematics anxiety," *Journal for Research in Mathematics Education, 21* (1990), 33–46. See also S. L. Beilock, L. A. Gunderson, G. Ramirez, and S. C. Levine, "Female teachers' math anxiety affects girls' math achievement," *Proceedings of the National Academy of Sciences, USA, 107* (2010), 1860–63.

2. L. Alexander and C. Martray, "The development of an abbreviated version of the Mathematics Anxiety Rating Scale," *Measurement and Evaluation in Counseling and Development 22* (1989), 143–50. Reprinted with permission.

3. *Foundations for Success: The Final Report of the National Mathematics Advisory Panel*, 2008, http://www2.ed.gov/about/bdscomm/list/mathpanel/report/final-report.pdf.

4. M. H. Ashcraft and E. P. Kirk, "The relationships among working memory, math anxiety, and performance," *Journal of Experimental Psychology: General, 130* (2001), 224–37.

5. C. M. Steele and J. Aronson, "Stereotype threat and the intellectual test performance of African-Americans," *Journal of Personality and Social Psychology, 69* (1995), 797–811.

6. J. Aronson et al., "When white men can't do math: Necessary and sufficient factors in stereotype threat," *Journal of Experimental Social Psychology, 35* (1999), 29–46.

7. E. E. Smith and J. Jonides, "Storage and executive processes in the frontal lobes," *Science, 283* (1999), 1657–61.

8. C. Rothmayr, et al., "Dissociation of neural correlates of verbal and non-verbal visual working memory with different delays," *Behavioral and Brain Functions, 3* (2007), 56. Redrawn with permission.

9. S. L. Beilock, R. J. Rydell, and A. R. McConnell, "Stereotype threat and working memory: Mechanisms, alleviation, and spill over," *Journal of Experimental Psychology: General, 136* (2007), 256–76.

10. M. S. DeCaro, K. E. Rotar, M. S. Kendra, and S. L. Beilock, "Diagnosing and alleviating the impact of performance pressure on mathematical problem solving," *Quarterly Journal of Experimental Psychology: Human Experimental Psychology* (2010) Forthcoming.

11. J. Wang et al., "Perfusion functional MRI reveals cerebral blood flow pattern under psychological stress," *PNAS, 102* (2005), 17804–809.

12. S. L. Beilock and T. H. Carr, "When high-powered people fail: Working memory and 'choking under pressure' in math," *Psychological Science, 16* (2005), 101–5.

13. D. Gimmig, P. Huguet, J. Caverni, and F. Cury, "Choking under pressure and working memory capacity: When performance pressure reduces fluid intelligence," *Psychonomic Bulletin & Review, 13* (2005), 1005–10.

14. J. C. Raven, J. E. Raven, and J. H. Court, *Progressive Matrices* (Oxford: Oxford Psychologists Press, 1998).

15. S. Hayes, C. Hirsh, and A. Mathews, "Restriction of working memory capacity during worry," *Journal of Abnormal Psychology, 17* (2008), 712–17.

16. D. G. Dutton and A. Aron, "Some evidence for heightened sexual attraction under conditions of high anxiety," *Journal of Personality and Social Psychology, 30* (1974), 510–17.

17. A. Mattarella-Micke et al., "Individual differences in math testing performance: Converging evidence from physiology and behavior," poster presented at the Annual Meeting of the Association for Psychological Science. Chicago, May 2008.

18. For a review, see S. F. Reardon, A. Atteberry, N. Arshan, and M. Kurlaender, (2009) "Effects of the California High School Exit Exam on Student Persistence, Achievement, and Graduation," Working Paper, Institute for Research on Education Policy & Practice, Stanford University, 2009–2012.

19. http://www.wonderlic.com/.

20. http://sports.espn.go.com/espn/page2/story?page=wonderlic/090218.

21. "Getting inside their heads," *Chicago Tribune,* February 20, 2008.
22. "Wondering about the Wonderlic?" *USA Today,* February 28, 2006.
23. College Board, *The Sixth Annual AP Report to the Nation,* http://www.collegeboard.com/apreport.

6. THE CHOKING CURE

1. *The Nation's Report Card,* National Assessment of Educational Progress, http://nces.ed.gov/nationsreportcard/.
2. C. Liston, B. S. McEwen, and B. J. Casey, "Psychosocial stress reversibly disrupts prefrontal processing and attentional control," *Proceedings of the National Academy of Sciences, USA, 106* (2009), 912–17.
3. C. Liston, B. S. McEwen, and B. J. Casey, "Psychosocial stress reversibly disrupts prefrontal processing and attentional control." *Proceedings of the National Academy of Sciences, USA, 106* (2009), 912–17. Redrawn with permission.
4. G. L. Cohen, J. Garcia, N. Apfel, and A. Master, "Reducing the racial achievement gap: A social-psychological intervention," *Science, 313* (2006), 1307–10.
5. G. L. Cohen et al., "Recursive processes in self-affirmation: Intervening to close the minority achievement gap," *Science, 324* (2009), 400–3. Note that this work included additional student cohorts that were not in the 2006 paper as well as additional self-affirmation writing boosters in the years following the initial intervention.
6. For a review, see S. L. Beilock, "Math performance in stressful situations," *Current Directions in Psychological Science, 17* (2008), 339–43.
7. G. Ramirez and S. L. Beilock, "The 'writing cure' as a solution to choking under pressure in math," Paper presented at the Annual Meeting of the Psychonomics Society. Chicago, November 2008.
8. J. W. Pennebaker, "Writing about emotional experiences as a therapeutic process," *Psychological Science, 8* (1997), 162–66. See also J. W. Pennebaker, *Writing to Heal: A Guided Journal for Recovering from Trauma and Emotional Upheaval* (Oakland: New Harbinger, 2004).
9. B. E. Depue, T. Curran, and M. T. Banich, "Prefrontal regions orchestrate suppression of emotional memories via a two-phase process," *Science, 317* (2009), 215–19. See also discussion of this paper, E. A. Holmes, M. L. Moulds, and D. Kavanagh, "Memory Suppression in PTSD Treatment?" *Science, 318* (2009), 1722, which argues that the proposal that suppression is a beneficial strategy for clinical intrusive memories is directly counter to treatment outcome data.
10. K. Klein and A. Boals, "Expressive writing can increase working memory capacity," *Journal of Experimental Psychology: General, 130* (2001), 520–33.
11. M. D. Lieberman et al., "Putting feelings into words: affect labeling disrupts amygdala activity in response to affective stimuli," *Psychological Science, 18* (2007), 421–28.
12. A. Lutz, H. A. Slagter, J. D. Dunne, and R. J. Davidson, "Attention regulation and monitoring in meditation," *Trends in Cognitive Science, 12* (2008), 163–69. See also "What the Beatles gave science," *Newsweek,* November 19, 2007.
13. G. Pagnoni et al., " 'Thinking about not-thinking': Neural correlates of conceptual processing during Zen meditation," *PLoS ONE,* e3083, 2008.

14. H. A. Slagter et al., "Mental training affects distribution of limited brain resources," *PLoS Biol. 5* e138, 2007.
15. S. L. Beilock, S. Todd, I. Lyons, and A. Lleras, "Meditating the pressure away," manuscript in progress.
16. "Wall Street bosses, Tiger Woods meditate to focus, stay calm," http://www.bloomberg.com/apps/news?pid=20601088&sid=aR2aP.X_Bflw&refer=muse#.
17. "Just say om," *Time*, August 4, 2003, http://www.time.com/time/magazine/article/0,9171,1005349,00.html
18. M. Shih, T. L. Pittinsky, and N. Ambady, "Stereotype susceptibility: Identity salience and shifts in quantitative performance," *Psychological Science, 10* (1999), 80–83.
19. D. M. Gresky, L. L. T. Eyck, C. G. Lord, and R. B. McIntyre, "Effects of salient multiple identities on women's performance under mathematical stereotypes," *Sex Roles, 53* (2005), 703–16.
20. For a review, see http://reducingstereotypethreat.org/reduce.html#deemphasizing. See also K. Danaher and C. S. Crandall, "Stereotype threat in applied settings re-examined," *Journal of Applied Social Psychology, 38* (2008), 1639–55.
21. D. M. Marx, S. J. Ko, and R. A. Friedman, "The 'Obama Effect': How a salient role model reduces race-based performance differences," *Journal of Experimental Social Psychology, 45* (2009), 953–56. Note that the lack of a significant difference in black and white test performance after the DNC acceptance speech was only seen in those people who actually watched the speech. See also J. Aronson, S. Jannone, M. McGlone, and T. Johnson-Campbell, "The Obama effect: An experimental test," *Journal of Experimental Social Psychology, 45* (2009), 957–60 for a discussion of limitations to the Marx et al. work.
22. N. Dasgupta and S. Asgari, "Seeing is believing: Exposure to counter stereotypic women leaders and its effect on the malleability of automatic gender stereotyping," *Journal of Experimental Social Psychology, 40* (2004), 642–58. Note that women's views were assessed using an Implicit Association Test (IAT) that was designed to assess the extent to which people automatically associated women with leadership qualities relative to supportive qualities.
23. S. E. Carrell, M. E. Page, and J. E. West, "Sex and science: How professor gender perpetuates the gender gap," National Bureau of Economic Research Working Paper, 14959, May 2009, http://www.nber.org/papers/w14959.
24. M. Johns, T. Schmader, and A. Martens, "Knowing is half the battle: Teaching stereotype threat as a means of improving women's math performance," *Psychological Science, 16* (2005), 175–79. This study deals with giving women information about gender math stereotypes, but also educating these women that the stereotypes need not apply to them.
25. National Science Foundation, Science and Engineering Indicators, 2006.
26. "Women in science: The battle moves to the trenches," *New York Times,* December 19, 2006.
27. "Cultivating Female Scientists," http://www.winchesterthurston.org, spring 2008.
28. The WISE program: http://www.gfs.org/academics/the-wise-program/index.aspx#1. Note that the program involves both male and female mentors.
29. American Society for Cell Biology, *Newsletter, 27*: 7 (2004).

7. CHOKING UNDER PRESSURE

1. See R. Jackson and S. L. Beilock, "Attention and performance," In D. Farrow, J. Baker, and C. MacMahon, eds., *Developing Elite Sports Performers: Lessons from Theory and Practice* (New York: Routledge, 2008), 104–18.
2. For a review, see S. L. Beilock and R. Gray, "Why do athletes 'choke' under pressure?" in G. Tenenbaum and R. C. Eklund, eds., *Handbook of Sport Psychology,* 3rd ed. (Hoboken, N.J.: Wiley, 2007), 425–44.
3. B. Calvo-Merino et al., "Action observation and acquired motor skills: An fmri study with expert dancers," *Cerebral Cortex, 15* (2005), 1243–49. See also B. Calvo-Merino et al., "Seeing or doing? Influence of visual and motor familiarity on action observation," *Current Biology, 16* (2006), 1905–10.
4. B. Calvo-Merino, D. E. Glaser, J. Grezes, R. E. Passingham, and P. Haggard, "Action observation and acquired motor skills: An fmri study with expert dancers," *Cerebral Cortex, 15* (2005), 1243–49. Redrawn with permission.
5. K. Yarrow, P. Brown, and J. W. Krakauer, "Inside the brain of an elite athlete: The neural processes that support high achievement in sports," *Nature Reviews Neuroscience, 10* (2009), 585–96. Mirror neurons, originally discovered in macaques, fire both during execution of goal-directed actions and during the observation of similar actions executed by another person.
6. R. E. Baumeister, "Choking under pressure: Self-consciousness and paradoxical effects of incentives on skillful performance," *Journal of Personality and Social Psychology, 46* (1984), 610–20. But, see, for a challenge and then discussion, B. R. Schlenker et al., "Championship pressures: Choking or triumphing in one's own territory?" *Journal of Personality and Social Psychology, 68* (1995), 632–41; R. E. Baumeister, "Disputing the effects of championship pressures and home audiences," *Journal of Personality and Social Psychology, 68* (1995), 644–48; B. R. Schlenker, S. T. Phillips, K. A. Boneicki, and D. R. Schlenker, "Where is the home choke?" *Journal of Personality and Social Psychology, 68* (1995), 649–52.
7. J. L. Butler and R. F. Baumeister, "The trouble with friendly faces: Skilled performance with a supportive audience," *Journal of Personality and Social Psychology, 75* (1998), 1213–30.
8. R. Jackson and S. L. Beilock, (2008) "Attention and performance," in D. Farrow, J. Baker, and C. MacMahon, eds., *Developing Elite Sports Performers: Lessons from Theory and Practice* (New York: Routledge, 2008), 104–18.
9. S. L. Beilock, T. H. Carr, C. MacMahon, and J. L. Starkes, "When paying attention becomes counterproductive: Impact of divided versus skill-focused attention on novice and experienced performance of sensorimotor skills," *Journal of Experimental Psychology: Applied, 8* (2002), 6–16.
10. R. Gray, "Attending to the execution of a complex sensorimotor skill: Expertise differences, choking and slumps," *Journal of Experimental Psychology: Applied, 10* (2004), 42–54.
11. N. A. Bernstein, *The Coordination and Regulation of Movements* (Oxford: Permagon, 1967).
12. D. V. Collins et al., "Examining anxiety associated changes in movement patterns," *International Journal of Sport Psychology, 32: 3* (2001), 223–42.
13. J. R. Pijpers, R. R. D. Oudejans, F. Holsheimer, and F. C. Bakker, "Anxiety-performance relationships in climbing: A process-oriented approach," *Psychology of Sport and Exercise, 4:3* (2003), 283–304.

14. J. G. Johnson and M. Raab, "Take the first: Option-generation and resulting choices," *Organizational Behavior and Human Decision Processes, 91* (2003), 215–29. For an additional discussion of optional decision-making without deliberation, see G. Gigerenzer, *Gut Feelings* (New York: Viking, 2007).

15. R. Poldrack et al., "The neural correlates of motor skill automaticity," *Journal of Neuroscience, 25* (2005), 5356–64.

16. T. Krigs et al., "Cortical activation patterns during complex motor tasks in piano players and control subjects: A functional magnetic resonance imaging study," *Neuroscience Letters, 278* (2000), 189–93.

17. For a review, see B. D. Hatfield and C. H. Hillman, "The psychophysiology of sport: A mechanistic understanding of the psychology of superior performance," in R. N. Singer, H. A. Hausenblas, and C. M. Janelle, eds., *Handbook of Sport Psychology* (New York: Wiley, 2001), 362–86. See also S. P. Deeny et al., "Electroencephalographic coherence during visuomotor performance: A comparison of cortico-cortical communication in experts and novices," *Journal of Motor Behavior, 41* (2008), 106–16.

18. J. Chen et al., "Effects of anxiety on EEG coherence during dart throw," Paper presented at the meeting of the 2005 World Congress, International Society for Sport Psychology, Sydney, Australia, August 2005.

19. J. T. Rietschel et al., "Electrocortical dynamics during competitive psychomotor performance," Paper presented at the Society for Neuroscience, Washington, D.C., 2008.

20. R. Jackson and S. L. Beilock, "Attention and performance," in D. Farrow, J. Baker, and C. MacMahon, eds., *Developing Elite Sports Performers* (New York: Routledge, 2008), 104–18.

21. R. S. W. Masters, R. C. J. Polman, and N. V. Hammond, " 'Reinvestment': A dimension of personality implicated in skill breakdown under pressure," *Personality and Individual Differences, 14* (1993), 655–66. Note that this is the original Reinvestment Scale. The Movement-Specific Reinvestment Scale is presented in chapter 2. Reprinted with permission.

22. G. Jordet, "When superstars flop: Public status and choking under pressure in international soccer penalty shootouts," *Journal of Applied Sport Psychology, 21* (2009), 125–30.

23. J. Stone, C. I. Lynch, M. Sjomeling, and J. M. Darley, "Stereotype threat effects on black and white athletic performance," *Journal of Personality and Social Psychology, 77* (1999), 1213–27.

24. S. L. Beilock et al., "On the causal mechanisms of stereotype threat: Can skills that don't rely heavily on working memory still be threatened?" *Personality & Social Psychology Bulletin, 32* (2006), 1059–71.

25. For a review, see A. M. Smith et al., "The 'yips' in golf: A continuum between a focal dystonia and choking," *Sports Medicine* 33 (2003), 13–31. See also C. M. Stinear et al., (2006). "The yips in golf: Multimodal evidence for two subtypes," *Medicine & Science in Sports and Exercise,* 1980–89.

26. D. A. Worthy, A. B. Markman, and W. T. Maddox, "Choking and excelling at the free throw line," *International Journal of Creativity & Problem Solving, 19* (2009), 53–58.

8. FIXING THE CRACKS IN SPORT AND OTHER FIELDS

1. "For free throws: 50 years of practice is no help," *New York Times,* March 4, 2009, http://www.nytimes.com/2009/03/04/sports/basketball/04freethrow.html.
2. "Elite learners under pressure," English Institute of Sport, http://www.eis2win.co.uk/pages/.
3. S. L. Beilock and T. H. Carr, "On the fragility of skilled performance: What governs choking under pressure?" *Journal of Experimental Psychology: General, 130* (2001), 701–25.
4. C. Y. Wan and G. F. Huon, "Performance degradation under pressure in music: An examination of attentional processes," *Psychology of Music, 33* (2005), 155–72.
5. E. H. McKinney and K. J. Davis, "Effects of deliberate practice on crisis decision performance," *Human Factors, 45* (2003), 436–44.
6. M. Jueptner et al., "Anatomy of motor learning: Frontal cortex and attention to action," *Journal of Neurophysiology, 77* (1997), 1313–24. See also K. Yarrow, P. Brown, and J. W. Krakauer, "Inside the brain of an elite athlete: The neural processes that support high achievement in sports," *Nature Reviews Neuroscience, 10* (2009), 585–96.
7. S. Berry, and C. Wood., "The cold-foot effect," *Chance 17* (2004), 47–51. The researchers looked at 2,003 kicks over the two seasons of which 1,565 were successful (78 percent success). There were 139 pressure kicks, and 101 (73 percent) were successful. When the defense called a timeout before the pressure kick (which happened on thirty-eight kicks), only twenty-four succeeded (about 63 percent).
8. D. M. Wegner, "How to think, say, or do precisely the worst thing for any occasion," *Science, 325* (2009), 48–51.
9. S. L. Beilock et al., "On the causal mechanisms of stereotype threat: Can skills that don't rely heavily on working memory still be threatened?" *Personality & Social Psychology Bulletin, 32* (2006), 1059–71.
10. C. Mesagno, D. Marchant, and T. Morris, "Alleviating choking: The sounds of distraction," *Journal of Applied Sport Psychology, 21* (2009), 131–47.
11. R. C. Jackson, K. J. Ashford, and G. Norsworthy, "Attentional focus, dispositional reinvestment and skilled motor performance under pressure," *Journal of Sport & Exercise Psychology, 28* (2006), 49–68.
12. D. F. Gucciardi and J. A. Dimmock, "Choking under pressure in sensorimotor skills: Conscious processing or depleted attentional resources?" *Psychology of Sport and Exercise, 9 (2008)*, 45–59.
13. G. Wulf and W. Prinz, "Directing attention to movement effects enhances learning: A review," *Psychonomic Bulletin and Review, 8* (2001), 648–60.
14. K. E. Flegal and M. C. Anderson, "Overthinking skilled motor performance: Or why those who teach can't do," *Psychonomic Bulletin & Review, 15* (2008), 927–32.
15. S. L. Beilock, S. A. Wierenga, and T. H. Carr, "Memory and expertise: What do experienced athletes remember?" in J. L. Starkes and K. A. Ericsson, eds., *Expert performance in sports: Advances in Research on Sport Expertise* (Champaign, Ill.: Human Kinetics, 2003), 295–320.
16. "Golfers take aim at victories with the help of a rifleman," *New York Times*, May 4, 2007.
17. H. Davis et al., "fmri bold signal changes in elite swimmers while viewing videos of personal failure," *Brain Imaging and Behavior, 2* (2008), 84–93. See also G. Miller, "Can Neuroscience Provide a Mental Edge?" *Science Magazine, 321* (2008).

18. P. Kruse et al., "β-Blockade used in precision sports: Effect on pistol shooting performance," *American Physiological Society* (1986), 417–20.
19. I. M. James, R. M. Pearson, D. N. W. Griffith, and P. Newbury, "Effect of Oxprenolol on stage-fright in musicians," *Lancet* (1977), 952–54.

9. CHOKING IN THE BUSINESS WORLD

1. J. D. Bransford and M. K. Johnson, "Contextual prerequisites for understanding: Some investigations of comprehension and recall," *Journal of Verbal Learning and Verbal Behavior, 11* (1972), 717–26.
2. For a review, see K. A. Paller and A. D. Wagner, "Observing the transformation of experience into memory," *Trends in Cognitive Science, 6* (2002), 93–102.
3. For a review, see P. M. Niedenthal, "Embodying emotion," *Science, 316* (2007), 1002–5.
4. T. L. Chartrand and J. A. Bargh, "The chameleon effect: The perception-behavior link and social interaction," *Journal of Personality and Social Psychology, 76* (1999), 893–910.
5. For a review, see P. M. Niedenthal, "Embodying emotion," *Science, 316* (2007), 1002–5. Note, at least in cultures where smiling is a frequent and accepted custom—even to strangers.
6. R. B. Zajonc, P. K. Adelmann, S. T. Murphy, and P. M. Niedenthal, "Convergence in the physical appearance of spouses," *Motivation and Emotion, 11* (1987), 335–46.
7. D. M. Wegner, "How to think, say, or do precisely the worst thing for any occasion," *Science, 325* (2009), 48–51.
8. S. L. Beilock, J. A. Afremow, A. L. Rabe, and T. H. Carr, " 'Don't miss!' The debilitating effects of suppressive imagery on golf putting performance," *Journal of Sport and Exercise Psychology, 23* (2001), 200–21.
9. F. C. Bakker, R. R. D. Oudejans, O. Binsch, and J. van der Kamp, "Penalty shooting and gaze behavior: Unwanted effects of the wish not to miss," *International Journal of Sport Psychology, 37* (2006), 265–80.
10. C. Kirschbaum, K. M. Pirke, and D. H. Hellhammer, "The 'Trier Social Stress Test'—a tool for investigating psychobiological stress responses in a laboratory setting," *Neuropsychobiology, 28* (1993), 76–81.
11. A. J. Fiocco, R. Joober, and S. J. Lupien, "Education modulates cortisol reactivity to the Trier Social Stress Test in middle-aged adults," *Psychoneuroendocrinology 32* (2007), 1158–63.
12. B. Ditzen et al., "Adult attachment and social support interact to reduce psychological but not cortisol responses to stress," *Journal of Psychosomatic Research, 64:5* (2008), 479–86.
13. C. Kirschbaum et al., "Sex-specific effects of social support on cortisol and subjective responses to acute psychological stress," *Psychosomatic Medicine, 57* (1995), 23–31.
14. For a review, see R. S. Weinberg and D. Gould, *Foundations of Sport & Exercise Psychology*, 4th ed. (Champaign, IL: Human Kinetics, 2007).
15. K. Klein and A. Boals, "Expressive writing can increase working memory capacity," *Journal of Experimental Psychology: General, 130* (2001), 520–33.
16. T. D. Wager et al., "Brain mediators of cardiovascular responses to social threat: Part I: Reciprocal dorsal and ventral sub-regions of the medial prefrontal cortex and heart-

rate reactivity," *Neuroimage, 47* (2009), 821–35; T. D. Wager et al., "Brain mediators of cardiovascular responses to social threat: Part II: Prefrontal-subcortical pathways and relationship with anxiety," *Neuroimage, 47* (2009), 836–51.

17. "Fear of sexual failure," http://www.4-men.org/sex-performance-anxiety.html.
18. For a review, see F. Hedon, "Anxiety and erectile dysfunction: A global approach to ED enhances results and quality of life," *International Journal of Impotence Research 15* (2003), S16–S19.
19. K. Ochsner and J. Gross, "Cognitive emotion regulation: Insights from social cognitive and affective neuroscience," *Current Directions in Psychological Science, 17* (2008), 153–58.
20. K. Deater-Deckard, M. D. Sewell, S. A. Petrill, and L. A. Thompson, (2010) "Maternal working memory and reactive negativity in parenting," *Psychological Science, 21*, 75–79.
21. C. S. Monk, E. B. McClure, E. E. Nelson et al., "Adolescent immaturity in attention-related brain engagement to emotional facial expressions," *NeuroImage, 20* (2003), 420–28. For a general overview, see also S. Choudhury, S. Blakemore, and T. Charman, "Social cognitive development during adolescence," *Social, Cognitive, and Affective Neuroscience, 1* (2006), 165–74.
22. M. S. DeCaro, K. E. Rotar, M. S. Kendra, and S. L. Beilock, "Diagnosing and alleviating the impact of performance pressure on mathematical problem solving," *Quarterly Journal of Experimental Psychology: Human Experimental Psychology* (2010).

EPILOGUE: ROMA NEVER FORGETS

1. S. L. Beilock, T. H. Carr, C. MacMahon, and J. L. Starkes, "When paying attention becomes counterproductive: Impact of divided versus skill-focused attention on novice and experienced performance of sensorimotor skills," *Journal of Experimental Psychology: Applied, 8* (2002), 6–16.
2. In addition to related citations from chapter 7, see also E. F. Wright and W. Jackson, "The home-course disadvantage in golf championships: Further evidence for the undermining effect," *Journal of Sport Behavior, 14* (1991), 51–61, and E. F. Wright and D. Voyer, "Supporting audiences and performance under pressure: The home-ice disadvantage in hockey," *Journal of Sport Behavior, 18* (1995), 21–29.
3. G. Jordet, "When superstars flop: Public status and choking under pressure in international soccer penalty shootouts," *Journal of Applied Sport Psychology, 21* (2009), 125–30.
4. D. A. Worthy, A. B. Markman, and W. T. Maddox, "Choking and excelling at the free throw line," *International Journal of Creativity & Problem Solving, 19* (2009), 53–58.

INDEX

ABOUT THE AUTHOR

Sian Beilock is one of the world's leading experts on the brain science behind "choking under pressure" and the many factors influencing all types of performance, from test-taking to public speaking to your golf swing. She is a tenured associate professor of psychology at the University of Chicago. She received a B.S. in cognitive science from the University of California, San Diego in 1997 and Ph.D.s in both kinesiology (sport psychology and motor learning) and psychology (cognitive psychology) from Michigan State University in 2003. Her research is supported by the National Science Foundation and the U.S. Department of Education.